SpringerBriefs in Applied Sciences and Technology

More information about this series at http://www.springer.com/series/8884

Filipe Manuel Clemente
Fernando Manuel Lourenço Martins
Rui Sousa Mendes

Social Network Analysis Applied to Team Sports Analysis

Filipe Manuel Clemente
Instituto Politécnico de Viana do Castelo,
    Escola Superior de Desporto e Lazer
Melgaço
Portugal

and

Instituto de Telecomunicações,
    Delegação da Covilhã
Covilhã
Portugal

Fernando Manuel Lourenço Martins
Instituto de Telecomunicações,
    Delegação da Covilhã
Covilhã
Portugal

and

Instituto Politécnico de Coimbra,
    Escola Superior de Educação
Coimbra
Portugal

Rui Sousa Mendes
Instituto Politécnico de Coimbra,
    Escola Superior de Educação
Coimbra
Portugal

ISSN 2191-530X   ISSN 2191-5318   (electronic)
SpringerBriefs in Applied Sciences and Technology
ISBN 978-3-319-25854-6   ISBN 978-3-319-25855-3   (eBook)
DOI 10.1007/978-3-319-25855-3

Library of Congress Control Number: 2015954989

Springer Cham Heidelberg New York Dordrecht London

Printed on acid-free paper

Springer International Publishing AG Switzerland is part of Springer Science+Business Media (www.springer.com)

Acknowledgments

The authors would like to thank Instituto Politécnico de Viana do Castelo—Escola superior de Melgaço, Instituto Politécnico de Coimbra—Escola Superior de Educação de Coimbra, and Instituto de Telecomunicações—Covilhã for the institutional support to make this book.

The authors would also to thank Dr. Dimitris Kalamaras and Prof. Frutuoso Silva for their scientific contribution in the previous works and for their suggestions for this book.

Finally, the authors would like to thank their families for the permanent support and for the patience with their scientific activity. For that reason, this book is dedicated to our families.

This work was supported by the FCT project PEst-OE/EEI/LA0008/2013.

Contents

About the Authors

Filipe Manuel Clemente is a Professor in Instituto Politécnico de Viana do Castelo, Escola Superior de Desporto e Lazer (Portugal), and post-doc researcher in Instituto de Telecomunicações, Delegação da Covilhã (Portugal). He has a Ph.D. in sport sciences—sports training in Faculty of Sport Sciences and Physical Education from University of Coimbra (Portugal). His research in sports training and sports medicine has led to more than 100 publications. He has conducted studies in computational tactical metrics, network analysis applied to team sports analysis, small-sided and conditioned games, physical activity, and health and sports medicine. He is a guest editor on Sports Performance and Exercise Collection in SpringerPlus Journal. He is also an editor in more three scientific journals. E-mail: Filipe.clemente5@gmail.com
    For further details see: http://www.researchgate.net/profile/Filipe_Clemente.

Fernando Manuel Lourenço Martins is a research member and scientific coordinator of Applied Mathematics group in Instituto de Telecomunicações, Delegação da Covilhã, Portugal, and a professor and course director of basic education in Instituto Politécnico de Coimbra, Escola Superior de Educação, Department of Education (Portugal). He has a Ph.D. in mathematics in University of Beira Interior (Portugal). His research in applied mathematics and statistical analysis has led to more than 110 publications. The research included early advances in network analysis applied to team sports analysis, statistical analysis in team sports, and mathematical teaching. He is a co-editor on Sports Performance and Exercise Collection in SpringerPlus Journal. E-mail: fmlmartins@ubi.pt
    For further details see: http://www.researchgate.net/profile/Fernando_Martins13.

Rui Sousa Mendes is a Dean and full-time Professor in Instituto Politécnico de Coimbra, Escola Superior de Educação, Department of Education (Portugal). He has a Ph.D. in Motor Learning and Control Faculty of Human Kinetics, University of Lisbon (Portugal). His research in motor control and learning and sports training has led to more than 120 publications. The research included early advances in dynamical systems applied to motor control and learning. He is a co-editor on Sports Performance and Exercise Collection in SpringerPlus Journal. E-mail: rmendes@esec.pt

For further details see: http://www.researchgate.net/profile/Rui_Mendes4.

Chapter 1
Introduction

Abstract Team sports games depend from the cooperation/interaction between teammates to avoid the opponent's strengths and exploit the opponent's weaknesses. The interaction between opposite players may also be considered in the specific field of team sports. Therefore, based on this dynamics that occurs in match it is possible to consider team sports as a cooperation-opposition game that depends from the interactions. To analyze this specific dynamics of connections it is possible to use techniques and methods based on social network analysis. Thus, the aim of this book is provide to the readers a summary of social network measures that can be applied to team sports and used to extract important information for match analysis interpretation.

Keyword Social network analysis · Team sports · Match analysis · Observation

1.1 Social Network Analysis in Team Sports

Team sports can be classified as systems (Davids et al. 2005). A system is said to be quasi-decomposable if it can be decomposed into quasi-isolated subsystems, with some interaction between them and with the context (Gréhaigne et al. 1997). In the system it is possible consider microsystems (few sub-systems with all their interactions) and infrasystems (few sub-systems with some of their interactions) (Gréhaigne et al. 1997). For this reason, these sub-systems may organize themselves into various types of networks either superimposed upon or merged inside the team (Gréhaigne et al. 1997).

Following the idea of subsystems that emerge in the match, a team of experts is not necessarily an expert team (Bourbousson et al. 2010). For that reason, the mechanism of synchronization and connection between teammates and their interaction with opposite teams have been analyzed in team sports analysis (McGarry 2005; Passos et al. 2011a, b; Duarte et al. 2012; Clemente et al. 2015a, b, c). This analyses allows to identify the properties of the teams and the dynamics that emerges

© The Author(s) 2016

F.M. Clemente et al., *Social Network Analysis Applied to Team Sports Analysis*,
SpringerBriefs in Applied Sciences and Technology,
DOI 10.1007/978-3-319-25855-3_1

during match (Travassos et al. 2013). Using such information it is possible to optimize the training plans and help to make decisions during matches (Clemente et al. 2014a, b).

In the last few years the social network analysis (SNA) have been occasionally applied to team sports analysis (Grund 2012; Peña and Touchette 2012; Duch et al. 2010; Clemente et al. 2015a, b, c). In the majority of the cases, the application of SNA to team sports analysis it was made in football and basketball (Grund 2012; Clemente et al. 2015a, b, c; Fewell et al. 2012; Cintia et al. 2015).

In the basketball analysis carried out in under-18 French players it was found small mutual forms of interactions and the results suggested that the coordination networks of the team were built on local coordination's chaining together in such a way as to not necessarily form a single unit (Bourbousson et al. 2010). In other hand a basketball analysis made in high-performance competition (NBA—first round of play-offs) it was observed a star pattern in the Bulls team who inbound only to the Point Guard at 60 % and for which most passes were between the Point Guard and other players (Fewell et al. 2012).

A singular analysis to water polo revealed that that the number of interactions between team members resulted in varied sub-systems coordination patterns of play, being possible to discriminate the successful and unsuccessful performance outcomes (Passos et al. 2011a, b).

In the case of football, one of first articles analysed the European Cup 2008 (Duch et al. 2010). In a comparison between ranks of best players in tournament, it was possible identify that the network centralities used by the authors were associated with expert opinions and ranks (Duch et al. 2010). In the specific case, Xavi (Spanish midfielder) was classified by network analysis as the best player. The same classification was made by the expert opinion of professionals.

The analysis to Spain (winner) during the last three matches in FIFA World Cup 2010 revealed that the clustering coefficient of the pass network increases with time, and stays high, indicating possession by Spanish players, which eventually leads to victory, even as the density of the pass network decreases with time (Cotta et al. 2013). In the previous tournament (FIFA World Cup 2006) it was found common and unique network dynamics of two competitive networks, compared with the large-scale networks that have previously been investigated in numerous works (Yamamoto and Yokoyama 2011).

Also in FIFA World Cup 2010 it was found that the greatest values of closeness centrality were associated with midfielders (Spain and Uruguay Teams) and defenders (Germany and Netherlands) teams (Peña and Touchette 2012). In the case of betweenness centrality it were found the greatest values in defenders (Spain team) and midfielders (Netherlands, Germany and Uruguay teams). Finally, the greatest values of PageRank centrality it were found in midfielders (Spain, Germany and Uruguay teams) and defenders (Netherlands team).

In a study that analysed a dataset of 283,259 passes in English Premier League of football (2006/2007 and 2007/2008 seasons) it was found that high levels of interaction (passes) lead to increased team performance (goals scored) (Grund 2012). In other hand, centralized tendencies lead to decreased team performance (Grund 2012).

The defense-attack transitions it was analysed in first Portuguese Football League (Malta and Travassos 2014). The results found that defensive midfielder was the prominent player in supported play. In other hand, centre forward was the prominent player in direct play. It was also found that the number of players surrounding the ball allow the emergence of the type of pass (short or long) (Malta and Travassos 2014). Also in the first Portuguese football league it was found that highest levels of scaled connectivity were associated with external defenders and midfielders during passing sequences (Clemente et al. 2014a). In the same study it was found that highest clustering coefficient was associated with midfielders and lowest values with goalkeeper and striker. A general analysis to digraphs in first Portuguese football league found that network density ranges between 0.2612 and 0.5846; network heterogeneity between 0.3900 and 0.4901; and network centralization between 0.1599 and 0.3182 (Clemente et al. 2015a, b, c).

In a study that analysed FIFA World Cup 2014 and Italian Serie A (2013/2014 season) the results found that performance of a team and network indicators are associated and may predict the outcomes of the games (Cintia et al. 2015). In a study that compared the national teams' performance during FIFA World Cup 2014 it were statistical differences between winning and losing teams in total links (greater values in winning teams) and density (greater values in winning teams) (Clemente et al. 2015a, b, c). It was also found statistical differences in total links and density between teams that reached the final and the teams that lose in round of 16. The comparison of performance variables with network properties found statistical positive correlations of goals scored, shots and shots on goal with total links, density and clustering coefficient (Clemente et al. 2015a, b, c).

Finally, an analysis of variance revealed statistical differences of prominent levels between tactical positions during FIFA World Cup 2014 (Clemente et al. 2015a, b, c). The results found that midfielders had the main values of in out-degree, in-degree, closeness and betweenness centralities in the majority of tactical line-ups.

Briefly, as possible of observe the studies of SNA in team sports have are recent. The majority of the analyses are centered in the passes between teammates are the range of measurements are not totally availed. For that reason, this book aims to introduce the possible measurements that can be used in match analysis process in team sports.

1.2 Structure of This Book

This book will start with the presentation of the main definitions and concepts of social network analysis. In this chapter, it will be possible to identify the main properties of networks and understand the network applications in the specific case of team sports analysis. After that, the following chapter will introduce the observational process to collect the data. In this chapter it will be possible to identify the main indicators to be analyzed, the main procedures to built the observational system and the process to generate the adjacency matrix to start the network analysis.

The presentation of network measurements and the sports interpretation will be organized in three chapters. The description begins with the network centralities that can be used to a micro-level of analysis. Following, the dependency measurements will be described to characterize the meso-level of analysis. Finally, the general measurements to characterize the macro-level of analysis will be introduced.

This book ends with a case-study to identify how is possible to use the social network analysis to analyze the patterns of play of a team.

References

Bourbousson, J., et al. (2010). Team coordination in basketball: Description of the cognitive connections among teammates. *Journal of Applied Sport Psychology, 22*(2), 150–166.

Cintia, P., Rinzivillo, S., & Pappalardo, L. (2015). A network-based approach to evaluate the performance of football teams. Machine Learning and Data Mining for Sports Analytics Workshop, Porto, Portugal.

Clemente, F. M., Couceiro, M. S., Martins, F. M. L., Mendes, R. S., et al. (2014a). Practical implementation of computational tactical metrics for the football game: Towards an augmenting perception of coaches and sport analysts. In Murgante et al. (Eds.), *Computational science and its applications* (pp. 712–727). New York: Springer.

Clemente, F. M., Couceiro, M. S., Martins, F. M. L., & Mendes, R. S. (2014b). Using network metrics to investigate football team players' connections: A pilot study. *Motriz, 20*(3), 262–271.

Clemente, F. M., Couceiro, M. S., et al. (2015a). Using network metrics in soccer: A macro-analysis. *Journal of Human Kinetics, 45*, 123–134.

Clemente, F. M., Martins, F. M. L., et al. (2015b). General network analysis of national soccer teams in FIFA World Cup 2014. *International Journal of Performance Analysis in Sport, 15* (1), 80–96.

Clemente, F. M., et al. (2015c). Midfielder as the prominent participant in the building attack: A network analysis of national teams in FIFA World Cup 2014. *International Journal of Performance Analysis in Sport, 15*(2), 704–722.

Cotta, C., et al. (2013). A network analysis of the 2010 FIFA world cup champion team play. *Journal of Systems Science and Complexity, 26*(1), 21–42.

Davids, K., Araújo, D., & Shuttleworth, R. (2005). Applications of dynamical systems theory to football. In T. Reilly, J. Cabri, & D. Araújo (Eds.), *Science and football V* (pp. 556–569). Oxon: Routledge Taylor & Francis Group.

Duarte, R., et al. (2012). Sports teams as superorganisms: Implications of sociobiological models of behaviour for research and practice in team sports performance analysis. *Sports Medicine, 42* (8), 633–642.

Duch, J., Waitzman, J. S., & Amaral, L. A. (2010). Quantifying the performance of individual players in a team activity. *PLoS ONE, 5*(6), e10937.

Fewell, J. H., et al. (2012). Basketball teams as strategic networks. *PLoS ONE, 7*(11), e47445.

Gréhaigne, J. F., Bouthier, D., & David, B. (1997). Dynamic-system analysis of opponent relationship in collective actions in football. *Journal of Sports Sciences, 15*(2), 137–149.

Grund, T. U. (2012). Network structure and team performance: The case of English Premier League soccer teams. *Social Networks, 34*(4), 682–690.

Malta, P., & Travassos, B. (2014). Characterization of the defense-attack transition of a soccer team. *Motricidade, 10*(1), 27–37.

McGarry, T. (2005). Soccer as a dynamical system: Some theoretical considerations. In T. Reilly, J. Cabri & D. Araújo (Eds.), *Science and football V* (pp. 570–579). London and New York: Routledge, Taylor & Francis Group.

Passos, P., et al. (2011a). Networks as a novel tool for studying team ball sports as complex social systems. *Journal of Science and Medicine in Sport, 14*(2), 170–176.

Passos, P., et al. (2011b). Tendencies of agents in team sports. *Journal of Motor Behavior, 43*(2), 155–163.

Peña, J. L., & Touchette, H. (2012). A network theory analysis of football strategies. In *arXiv preprint arXiv.* p. 1206.6904.

Travassos, B., et al. (2013). Performance analysis in team sports: Advances from an ecological dynamics approach. *International Journal of Performance Analysis in Sport, 13*(1), 83–95.

Yamamoto, Y., & Yokoyama, K. (2011). Common and unique network dynamics in football games. *PLoS ONE, 6*(12), e29638.

Chapter 2
Social Network Analysis: Concepts and Definitions

Abstract The aim of this chapter is to present the main definitions and concepts associated with social network analysis. These definitions and concepts will help to understand the fundaments of graph theory and the following micro, meso- and macro-measurements.

Keywords Social network analysis · Concepts · Graph theory · Digraphs

2.1 Definitions and Concepts

The Social Network Analysis (SNA) is based on Graph Theory (Barnes and Harary 1983), a mathematical study of sets of vertices connected by edges. The techniques model pairwise relations between the vertices.

To better understand the network perspective, consider the social network of Team of football players shown in Fig. 2.1. It is an example of a sociogram, also called a network graph, which is a common way of visualizing networks. Like all networks, it consists of two primary building blocks: vertices (also called nodes or points or agents) and edges (also called lines or ties or arcs or connections). The vertices are represented by number of players of a Team, and the edges are represented by the lines that point from one vertex to another. Other similar context, for example, is presented in (Hansen et al. 2011) to Twitter's users.

The kind of relations considered between vertices in graph theory, in terms of mathematics, are described by graphs that represent networks and those graphs are called directed graphs, undirected graphs, and weighted graphs (digraphs) (Pavlopoulos et al. 2011). Thus, in the following presented some elementary concepts on Graphs Theory used in SNA.

© The Author(s) 2016

F.M. Clemente et al., *Social Network Analysis Applied to Team Sports Analysis*,
SpringerBriefs in Applied Sciences and Technology,
DOI 10.1007/978-3-319-25855-3_2

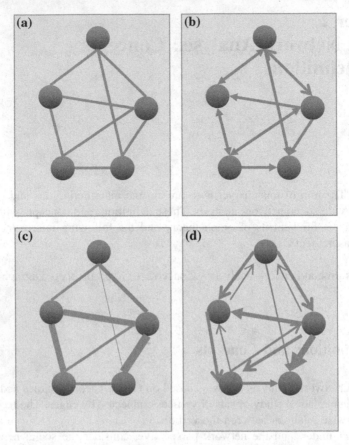

Fig. 2.1 Examples of unweighted graph (**a**), unweighted digraph (**b**), weighted graph (**c**), and weighted digraph (**d**)

Definition 2.1 (Gross and Yellen 2004) A graph $G = (V, E)$ consists of two sets V and E, where V is a set of vertices (or nodes), E is a set of edges and each edge has a set of one or two vertices associated to it, which are called its endpoints (or neighbors) and an edge is said to joint its endpoints.

Remark 2.1 Given two vertices u and v. Then the edge, the single connection between vertices u and v, is represented by pair (u, v).

Definition 2.2 (Gross and Yellen 2004) If vertex u is an endpoint of edge l, then u is said to be incident on l, and l is incident on u.

Definition 2.3 (Gross and Yellen 2004) A vertex u is adjacent to vertex v if they are joined by any edge.

Definition 2.4 (Pavlopoulos et al. 2011) A complete graph is a graph in which every pair of vertices is adjacent.

Definition 2.5 (Gross and Yellen 2004) Two adjacent vertices may be called neighbors.

Example 2.1 Let $G = (V, E)$ be a unweighted graph with $n = 5$ vertices (Fig. 2.2).

Definition 2.6 (Pavlopoulos et al. 2011) An undirected graph is connected if one can get from any vertex to any other vertex by following a sequence of edges.

Definition 2.7 (Gross and Yellen 2004) A walk in a graph G is an sequence of vertices and edges $W = v_0, e_1, v_1, e_2, \ldots, e_n, v_n$ shut that for $j = 1, \ldots, n$, the vertices v_{j-1} and v_j are the endpoints of the edge e_j.

Definition 2.8 (Gross and Yellen 2004) A graph that has no loops and includes no more than one edge between a pair of vertices is called a simple graph.

Definition 2.9 (Wasserman and Faust 1994) A graph G_S is a subgraph of G if the set of the vertices of G_S is a subset of vertices of G, and the set of edges in G_S is a subset of the edges in the graph G.

Definition 2.10 (Wasserman and Faust 1994) A subgraph G_S, is generated by a set of vertices if G_S has vertex set and edge set where the set of edges includes all edges of graph G that are between pairs of vertices of the G_S.

Remark 2.2 (Wasserman and Faust 1994) In Definition 2.10 is define vertex-generated subgraphs. In SNA context is only a subset of the g members of network.

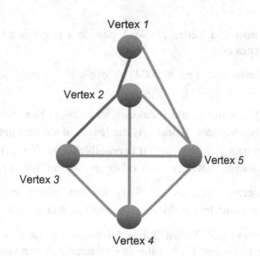

Fig. 2.2 Unweight graph G with five vertices. The set of all vertices is $V = \{n_1, n_2, n_3, n_4, n_5\}$ where $n_i := Vertex\,i$. The edge $e_1 = (n_1, n_5)$ *is the single connection between vertices n_1 and n_5. The set all edges is* $E = \{e_1, e_2, e_3, e_4, e_5\} = \{(n_1, n_5), (n_1, n_2), (n_2, n_3), (n_2, n_5), (n_2, n_4), (n_3, n_5)\}$. *The vertex n_1 is neighbor of vertex n_5 and the set of all neighbors of n_5 is* $N(n_5) = \{n_1, n_2, n_3, n_4\}$

Definition 2.11 (Wasserman and Faust 1994) A dyad, representing a pair of agents and the possible edge between them is a (vertex-generated) subgraph consisting of a pair of vertices and the possible edge between the vertices.

Remark 2.3 (Wasserman and Faust 1994) In a graph an unordered pair of vertices can be in only one of two states: either two vertices are adjacent or they are not adjacent.

Definition 2.12 (Wasserman and Glaskiewicz 1994) A triad is a subgraph consisting of the tree vertices and the possible edges among them.

Remark 2.4 (Wasserman and Faust 1994) In a graph, a triad may be in one of four possible sates, depending on whether, zero, one, two or three edges are presented among of tree vertices in the triad.

Definition 2.13 (Wasserman and Faust 1994) A triad involving the vertices u, v, z is transitive if $(u, v), (v, z) \in E$ then $(u, z) \in E$.

Theorem 2.1 (Wasserman and Faust 1994) *A graph is transitive if every triad it contains is transitive.*

Definition 2.14 (Gross and Yellen 2004) The length of walk is the number of edges (counting repetitions).

Definition 2.15 (Gross and Yellen 2004) A walk is closed if the initial vertex is also the final vertex; otherwise, it is open.

Definition 2.16 (Gross and Yellen 2004) A trail in a graph is a walk shut that no edge occurs than more once.

Definition 2.17 (Gross and Yellen 2004) A path in a graph is a trial shut that no internal vertex is repeated.

Definition 2.18 (Gross and Yellen 2004) A cycle is a closed path of length at least 1.

Definition 2.19 (Wasserman and Glaskiewicz 1994) The geodesic distance, $d(n_i, n_j)$, between two vertices, n_i and n_j is the length of shortest path between them and in cases that no path was generated it is possible to set $d(n_i, n_j) = \infty$ assuming that the vertices are so far between each other so they are not connected.

Remark 2.5 (Wasserman and Faust 1994) In unweighted graph the distance between n_i and n_j is equal to the distance between n_j and n_i; $d(n_i, n_j) = d(n_j, n_i)$.

Definition 2.20 (Gross and Yellen 2004) Given two graphs $G_1 = (V_1, E_1)$ and $G_2 = (V_2, E_2)$, where G_1 and G_2 has the same number of vertices. G_1 and G_2 are isomorphic if for all $n_i, n_j \in V_1$ and $v_i, v_j \in V_2$ there exists a one-to-one mapping $f : V_1 \rightarrow V_2$, $f(n_i) = v_i$ and $f(n_j) = v_j$, such that $(n_i, n_j) \in E_1$ if only if $(v_i, v_j) \in E_2$.

Definition 2.21 (Pavlopoulos et al. 2011) Bipartite graph is an undirected graph $G = (V, E)$ in which V can be partitioned into two sets V_1 and V_2 such that $(u, v) \in E$ implies either $u \in V_1$ and $v \in V_2$ or $v \in V_1$ and $u \in V_2$.

The Definitions 2.1–2.21 represent the class of graphs called of undirected and unweighted graph.

The other broad class are the unweighted directed graphs that we show the following in Definition 2.22–2.41.

Definition 2.22 (Gross and Yellen 2004) A directed graph (or digraph) is a graph each of whose edges is directed shut that a direct edge (or arc) is an edge that linked in the initial vertex to the terminal vertex.

Remark 2.6 (Wasserman and Faust 1994) The difference between an arc (in a digraph) and a edge (in a graph) is that an arc is an ordered pair of vertices (to reflect the direction of the edge between of two vertices) whereas a edge is unordered pair of vertices.

Definition 2.23 (Wasserman and Glaskiewicz 1994) Considering that a given vertex is the first (sender) or second (receiver) in the ordered pair defining the arc. A vertex u is adjacent to vertex v if the arc between u and v exist in digraph and a vertex v is adjacent to vertex u if the arc between v and u exist in digraph.

Remark 2.7 (Wasserman and Faust 1994) When a digraph is presented as a diagram the vertices as represented as points and the arcs represented as directed arrows. The arc (u, v) is represented by any arrow from the point representing u to the point representing v.

Example 2.2 Let $G = (V, E)$ be a unweighted digraph with $n = 5$ vertices (Fig. 2.3).

Definition 2.24 (Gross and Yellen 2004) A digraph that has no loops and includes no more than one arc that linked in the initial node to the terminal node is called a simple digraph.

Definition 2.25 (Wasserman and Faust 1994) A dyad in digraphs is subgraph that consisting of the two vertices and the possible arcs between of them.

Definition 2.26 (Wasserman and Faust 1994) A triad in digraphs is subgraph that consisting of the tree vertices and the possible arcs between of them.

Theorem 2.2 (Wasserman and Faust 1994) A digraph is transitive if every triad it contains is transitive.

Definition 2.27 (Gross and Yellen 2004) In a digraph G, W is directed walk, if the edge e_j is directed from v_{j-1} to v_j, where v_{j-1} and v_j are vertices of the G.

Definition 2.28 (Wasserman and Faust 1994) The length of directed walk is the number of instances arcs in it (an arc is counted each time it occurs in the walk).

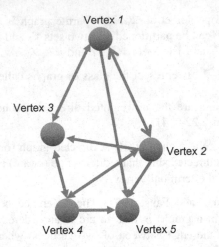

Fig. 2.3 Unweight digraph with five vertices. The set of all vertices is $V = \{n_1, n_2, n_3, n_4, n_5\}$ where $n_i := Vertex\,i$. The arc $e_1 = (n_2, n_5)$ is the single connection of n_2 for n_5. The set of all arcs is $E = \{e_1, e_2, e_3, e_4, e_5, e_6, e_7, e_8, e_9, e_{10}, e_{11}, e_{12}\} = \{(n_2, n_5), (n_2, n_3), (n_2, n_4), (n_2, n_1), (n_1, n_2),$ $(n_4, n_2), (n_4, n_3), (n_4, n_5), (n_3, n_4), (n_3, n_1), (n_1, n_3), (n_1, n_5)\}$. The vertex n_5 is outgoing neighbor of vertex n_1 and the set of all outgoing neighbors of n_1 is $N^{out}(n_1) = \{n_2, n_3, n_5\}$. n_2 is incoming neighbor of n_1 and he set of all incoming neighbors of n_1 is $N^{in}(n_1) = \{n_2, n_3\}$

Definition 2.29 (Wasserman and Glaskiewicz 1994) A directed trail in a digraph is a directed walk in which no arc is included more than one.

Definition 2.30 (Bondy and Murty 2008) A directed path (or simply a path) in digraph is a directed walk in which no vertex and no arc is included more than once.

Definition 2.31 (Wasserman and Faust 1994) A length of a path is the number of arcs in it.

Definition 2.32 (Wasserman and Glaskiewicz 1994) A semiwalk joining vertices v_i and v_j is a sequence of vertices and arcs which successive pairs of vertices are incident with an arc from the first to the second, or by an are from the second to the first.

Remark 2.8 In a semiwalk the direction of the arcs is irrelevant.

Definition 2.33 (Wasserman and Faust 1994) The length of a semiwalk is the number of instances of arcs in it.

Definition 2.34 (Wasserman and Faust 1994) A semipath joining vertices v_i and v_j is a sequence of distinct vertices were all successive pairs of vertices are connected by an arc from the first to the second, or by an arc from the second to the first for all successive pairs of vertices.

Remark 2.9 In a semipath the direction of the arcs is irrelevant.

Definition 2.35 (Wasserman and Faust 1994) The length of a semipath is the number of instances of arcs in it.

Definition 2.36 (Wasserman and Faust 1994) A cycle in digraph is a closed direct walk of at least tree vertices in which all vertices except the first and the last are distinct.

Definition 2.37 (Wasserman and Faust 1994) A semicycle in digraph is a closed direct semiwalk of at least tree vertices in which all vertices except the first and the last are distinct.

Remark 2.10 (Wasserman and Faust 1994) In a semicycle the arcs may go in either direction, whereas in a cycle the arcs must "point" in same direction.

Definition 2.38 (Wasserman and Glaskiewicz 1994) Given a pair of vertices n_i, n_j of digraph G with n vertices, $i, j = 1, \ldots, n$ and $i \neq j$. A pair of vertices n_i and n_j, is:

(i) Weakly connected if they are joined by a semipath;
(ii) Unilaterally connected if they are joined by a path from n_i, n_j, or a path from n_i to n_j;
(iii) Strongly connected if there is a path from n_i to n_j and a path from n_j to n_i; the path from n_i to n_j may contain different vertices and arcs than the path from n_j to n_i;
(iv) Recursively connected if they are strongly connected and the path from n_i to n_j uses the same vertices and arcs as the path from n_j to n_i in reverse order.

Definition 2.39 (Gross and Yellen 2004) A directed graph is:

(i) Weakly connected if all pairs of vertices are Weakly connected;
(ii) Unilaterally connected if all pairs of vertices are unilaterally connected;
(iii) Strongly connected if all pairs of vertices are strongly connected;
(iv) Recursively connected if all pairs of nodes are recursively connected.

Definition 2.40 (Rubinov and Sporns 2010; Wasserman and Faust 1994) The geodesic distance, $d(n_i, n_j)$, between two vertices, n_i and n_j is the length of shortest path between them and in cases that no path was generated it is possible to set $d(n_i, n_j) = \infty$ assuming that the vertices are so far between each other so they are not connected.

Remark 2.11 In unweighted digraph the distance between n_i and n_j is not equal to the distance between n_j and n_i; $d(n_i, n_j) \neq d(n_j, n_i)$ (Wasserman and Faust 1994).

Definition 2.41 (Chen 1997) Two digraphs are said to be isomorphic if verify simultaneously the following conditions:

(i) Their associated undirected graphs are isomorphic;
(ii) The directions of their corresponding edges are preserved for some correspondences of (i).

Definition 2.42 (Wasserman and Glaskiewicz 1994) In the case of a any graph (digraph) with all connections measured between any two vertices is called weighted graph (digraph).

Remark 2.12 (Wasserman and Faust 1994) The value of the weight of edge between any two vertices, v_i, v_j is represented by w_{ij}. In a valued graph the edge between vertex u and vertex v is identical to the edge between vertex v and vertex u and thus there is only a single value for each unordered pair of vertices. In the case weighted digraph, a valued digraph the arc between vertex u and vertex v no is identical to the arc between vertex v and vertex u and thus there is two values, one for each possible arc for the ordered pair of vertices.

Example 2.3 Given one weighted graph G_1 and one weighted digraph G_2, both with $n = 5$ vertices (Fig. 2.4).

Definition 2.43 (Wasserman and Faust 1994) A dyad in weighted graphs has a edge between vertices with a specific strength and in weighted digraphs has arcs between vertices with a specific strength, respectively.

Definition 2.44 (Wasserman and Faust 1994) A triad in weighted graphs has a edges between vertices with a specific strength, respectively, and in weighted digraphs has arcs between vertices with a specific strength, respectively.

Definition 2.45 (Umeyama 1988) Two weighted undirected graphs, G_1, G_2 are said to be isomorphic if verify simultaneously the following conditions:

(i) Their associated unweighted and undirected graphs are isomorphic;

(ii) $\sum_{i=1}^{n} \sum_{j=1}^{n} \left(w_{ij}^1 - w_{ij}^2 \right)^2 = 0$, where w_{ij}^1 is the weight of the edge between any two vertices, n_i, n_j and w_{ij}^2 is the weight of the edge between any two vertices, v_i, v_j with $f(n_i) = v_i$ and $f(n_j) = v_j$, such that $(n_i, n_j) \in E_1$ and $(v_i, v_j) \in E_2$.

Definition 2.46 (Bunimovich and Webb 2014) Two weighted digraphs, G_1, G_2 are said to be isomorphic if for all $n_i, n_j \in V_1$ and $v_i, v_j \in V_2$ there exists a one-to-one mapping $f : V_1 \rightarrow V_2$, $f(n_i) = v_i$ and $f(n_j) = v_j$, such that w_{ij}^1 is the weight of the arc from n_i to n_j with $(n_i, n_j) \in E_1$ if only if w_{ij}^2 is the weight of the arc from v_i to, v_j with $(v_i, v_j) \in E_2$ and $w_{ij}^1 = w_{ij}^2$.

The information of an unweighted or weighted graph or digraph may also be expressed in a variety of ways in matrix form presented in the Definition 16.

Definition 2.47 (Wasserman and Faust 1994) The $n \times n$ matrix $A = [a_{ij}]$ is the adjacency matrix of a unweighted graph or unweighted digraph with n vertices such that $a_{ij} = 1$ if $(n_i, n_j) \in E$ or $a_{ij} = 0$ if $(n_i, n_j) \notin E$, $i, j = 1, \ldots, n$. In the case of weighted graphs or weighted digraphs $a_{ij} = w_{ij}$ if $(n_i, n_j) \in E$ or $a_{ij} = 0$ otherwise. Thus, the matrix A is called weighted adjacency matrix.

(a)

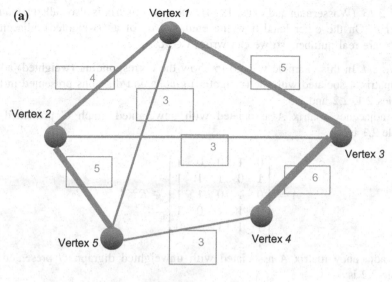

(b)

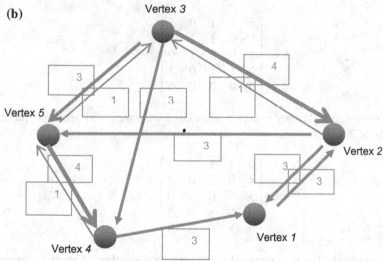

Fig. 2.4 Weighted graph (**a**) and a weighted digraph (**b**). **Weighted graph G_1.** The set of all vertices is $V_1 = \{n_1, n_2, n_3, n_4, n_5\}$ where $n_i := Vertex\,i$. The edge $e_1 = (n_1, n_5)$ *is the single connection between vertices* n_1 *and* n_5. *The set all edges is* $E = \{(n_1, n_5), (n_1, n_2), (n_1, n_3), (n_2, n_3), (n_2, n_5), (n_5, n_4), (n_4, n_3)\}$. *The vertex* n_1 *is neighbor of vertex* n_2 *and the set of all neighbors of* n_2 *is* $N(n_2) = \{n_1, n_3, n_5\}$. *The* $w_{12} = 4$ *is the weight of the edge* $(n_1, n_2) \in E_1$. *We can observe that* $w_{12} = w_{21} = 4$. **Weighted digraph G_2.** The set of all vertices is $V_2 = \{n_1, n_2, n_3, n_4, n_5\}$ where $n_i := Vertex\,i$. The arc $e_5 = (n_3, n_2)$ *is the single connection of n_3 for n_2 and the* $w_{32} = 4$ *is the weight of the arc* $(n_3, n_2) \in E_2$. *However,* $w_{23} = 1$ *and, in general,* $w_{ij} \neq w_{ji}$. The set of all arcs is $E = \{e_1, e_2, e_3, e_4, e_5, e_6, e_7, e_8, e_9, e_{10}, e_{11}\} = \{(n_1, n_2), (n_2, n_1), (n_2, n_5), (n_2, n_3), (n_3, n_2), (n_3, n_4), (n_3, n_5), (n_5, n_3), (n_5, n_4), (n_4, n_5), (n_4, n_1)\}$. The vertex n_3 *is outgoing neighbor of vertex* n_2 *and the set of all outgoing neighbors of* n_2 *is* $N^{out}(n_2) = \{n_1, n_3, n_5\}$. n_1 *is incoming neighbor of* n_2 *and he set of all incoming neighbors of* n_2 *is* $N^{in}(n_2) = \{n_1, n_3\}$

Remark 2.13 (Wasserman and Faust 1994) One $n \times n$ matrix is also called a matrix of order n. On the other hand how the elements a_{ij} of an (weighted) adjacency matrix A are real numbers so we can write $A \in \mathcal{R}^n$.

Example 2.4 In this example will go to show the corresponding (weighted) adjacency matrix associated with the (weighted) graphs and digraphs presented in the Examples 2.1, 2.2 and 2.3.

The adjacency matrix A associated with unweighted graph G presented in Example 2.1 is

$$A = \begin{bmatrix} 0 & 1 & 0 & 0 & 1 \\ 1 & 0 & 1 & 1 & 1 \\ 0 & 1 & 0 & 1 & 1 \\ 0 & 1 & 1 & 0 & 1 \\ 1 & 1 & 1 & 1 & 0 \end{bmatrix} \in \mathcal{R}^{5 \times 5}.$$

The adjacency matrix A associated with unweighted digraph G presented in Example 2.2 is

$$A = \begin{bmatrix} 0 & 1 & 1 & 0 & 1 \\ 1 & 0 & 1 & 1 & 1 \\ 1 & 0 & 0 & 1 & 0 \\ 0 & 1 & 1 & 0 & 1 \\ 0 & 0 & 0 & 0 & 0 \end{bmatrix} \in \mathcal{R}^{5 \times 5}.$$

The weighted adjacency matrix A associated with weighted graph G_1 presented in Example 2.3(a) is

$$A = \begin{bmatrix} 0 & 4 & 5 & 0 & 3 \\ 4 & 0 & 3 & 0 & 5 \\ 5 & 3 & 0 & 6 & 0 \\ 0 & 0 & 6 & 0 & 3 \\ 3 & 5 & 0 & 3 & 0 \end{bmatrix} \in \mathcal{R}^{5 \times 5}.$$

The weighted adjacency matrix A associated with weighted digraph G_2 presented in Example 2.3(b) is

$$A = \begin{bmatrix} 0 & 3 & 0 & 0 & 0 \\ 3 & 0 & 1 & 0 & 3 \\ 0 & 4 & 0 & 3 & 3 \\ 3 & 0 & 0 & 0 & 1 \\ 0 & 0 & 1 & 4 & 0 \end{bmatrix} \in \mathcal{R}^{5 \times 5}.$$

References

Barnes, J. A., & Harary, F. (1983). Graph theory in network analysis. *Social Networks, 5*, 235–244.

Bondy, J. A., & Murty, U. S. R. (2008). *Graph Theory*. USA: Springer.

Bunimovich, L., & Webb, B. (2014). *Isospectral Transformations: A New Approach to Analysing Multidimensional Systems and Networks*. New York, USA: Springer.

Chen, W. K. (1997). *Graph Theory and Its Engineering Applications*. Singapore: World Scientific Publishing Co., Pte. Ltd.

Gross, L., & Yellen, J. (2004). *Handbook of Graph Theory*. New York, USA: CRC Press.

Hansen, D., Shneiderman, B., & Smith, M. A. (2011). *Analyzing Social Media Networks with NodeXL: Insights from a Connected World*. Burlington, USA: Morgan Kaufmann—Elsevier.

Pavlopoulos, G. A., et al. (2011). Using graph theory to analyze biological networks. *BioData Mining, 4*(1), 10.

Rubinov, M., & Sporns, O. (2010). Complex network measures of brain connectivity: Uses and interpretations. *NeuroImage, 52*(3), 1059–1069.

Umeyama, S. (1988). An eigendecomposition approach to weighted graph matching problems. *IEE Transactions on Pattern Analysis and Machine Intelligence, 10*(5), 1988.

Wasserman, S., & Faust, K. (1994). *Social Network Analysis: Methods and Applications*. New York, USA: Cambridge University Press.

Wasserman, S., & Glaskiewicz, J. (1994). *Advances in Social Network Analysis: Research in the Social and Behavioral*. California, USA: SAGE Publications.

Chapter 3
Observational Tools to Collect Data in Team Sports

Abstract Team sports lead to permanent interactions between teammates. For that reason, the specific structure of interactions and team's dynamics must be carefully analysed in order to improve the sports training and strategy used for the match. For that reason, the social network analysis has been used in the last few years to identify the properties of graphs and to measure the centrality levels of players and tactical positions in the collective organization and dynamics of team sports. Nevertheless, the match analysis based on social network analysis must follow some specific requirements. Therefore, this chapter aims to describe the required observational procedures for network analysis in team sports and to show some software to process the analysis and to extract the data to measure the network properties.

Keywords Team sports · Match analysis · Social network analysis · Observation

3.1 The Collective Nature of Team Sports

As in any organization or collective structure the team sports can be described as a group of players that works together and in synchronism to achieve common goals (Lusher et al. 2010). Therefore, concepts such as cohesiveness and hierarchies are also presented in team sports and can be considered decisive to determine the dynamics in match (Grund 2012). Nevertheless, to achieve a high-level of collective organization and teammates' organization during match it is required a consistent capacity of inter-players synchronization in order to act as a whole to avoid the strengths and to exploit the weaknesses of the opponent (Gréhaigne et al. 1997). For that reason, it is required more than a sum of good players. To act as a whole the players must act as one (Grund 2012). Thus, it is possible to consider that a team of experts it is not necessarily an expert team (Bourbousson et al. 2010).

© The Author(s) 2016   19
F.M. Clemente et al., *Social Network Analysis Applied to Team Sports Analysis*,
SpringerBriefs in Applied Sciences and Technology,
DOI 10.1007/978-3-319-25855-3_3

Another interesting evidence in team sports as in any collective organization it is the sub-groups that emerges from the natural interactions processes (Vilar et al. 2013). These sub-groups (or clusters) emerge constrained by the specific dynamics of the game (Bourbousson et al. 2010). In the specific case of football it is possible to identify specific clusters of interaction between forward, playmaker and external midfielder during attacking plays. In other hand, considering the defensive coverage as one linkage indicator, central-defenders and external defenders may constitute a specific cluster that occurs in the match (Duarte et al. 2012). In the case of handball will be normal occurring specific clusters between right wing and right back. In all these cases, the spatio-temporal relationship and the specific tactical positions and missions determines the occurrence of sub-graphs inside the team. Moreover, the model of playing may also determine these sub-graphs. In the case of the FIFA World Cup 2014, it was verified that Germany (champion of this tournament) had the greater values of density and clustering coefficient, revealing that the small occurrence of the sub-graphs was associated with the best performances in the tournament (Clemente et al. 2015a, b, c, d). In fact, the tendency to play as a whole may be associated with best performances in the generality of competitions as verified in previous studies in English Premier League of football (Grund 2012). The style of play it is also a factor that determine the specific patterns of interactions. Teams that play using ball circulation and the involvement of all players to build the attack, increases the density and homogeneity of participation, thus reducing the occurrences of sub-graphs. By the other hand, teams that prioritize the faster counter-attack and attacking transitions tend to generate more clusters (Clemente et al. 2015a, b, c, d).

Besides the interactions that occur between teammates, it is also important to consider the relationship with opponents. This rapport of strength occur in any space of the field and it is more or less variables based on the specific strategy used by the teams and the rules of the sport (Gréhaigne et al. 1999, 2011).

As possible to observe the dynamic of the game may determine the patterns of interactions (McGarry 2005). Nevertheless, the sports training may provide the major contribution to determine the interactions. For that reason, variables such as model of play, tactical line-up, tactical missions and others that are associated with coaches' intervention may be determinants to constrain the collective organization and flexibility of team play. Therefore, the understanding of these variables will help to optimize the team's performance. Following this idea, the strength of match analysis in sports have been growing in the last decade (Coutts 2014). The process of match analysis consists into collect the most important variables of play of a team and process the data to generate a final report that will help to identify the strengths and weaknesses of a team (Carling et al. 2005). The characterization of specific behaviors and collective/individual patterns will be also collected, thus providing important information to make decisions about the training planning and the strategy to use per each match (Barreira et al. 2013).

Briefly, the coach will use the match analysis as a permanent cycle of feedback to optimize the sports training and the decisions during matches. Nevertheless, there are multiple possibilities to observe, process and report the data. In practical contexts, the head-coach determines the most important variables and information that want to use. However, in the practical context there are lacks in proper software to process the data in a integrative point-of-view (Clemente et al. 2014a, b). The majority of the systems depends from the human operator and minimizes the possibilities of analysis to specific notational variables such as passes, recoveries, dribble or shots (Hughes and Bartlett 2002; Franks and McGarry 1996). In other hand, the majority of the computational software uses the automatic tracking systems only to analyze the time-motion variables and to provide information about the activity profile of the players (Carling et al. 2008). Despite of this current tendency in match analysis in practical contexts, the scientific community has been developing different approaches to optimize the data collecting and to use the exact sciences to process the data and report the major evidences in a quickly and user-friendly fashion (Travassos et al. 2013; Clemente et al. 2014a, b; Bartlett et al. 2012). Therefore, the next section will describe the recent computational and mathematical methods that have been proposed in the scientific community dedicated to the field of match analysis.

3.2 Match Analysis in Team Sports: Contribution for Sports Training

As previously described the match analysis can be carried out based on different approaches, techniques and methods. In this case, it is possible to categorize the analysis in (Clemente et al. 2014a, b): (i) traditional notational analysis based on individual variables and categories of actions; (ii) observational processes that collect information about tactical behaviors and collective organization; (iii) notational analysis based on computation that uses the categories and observational variables to establish some patterns of actions; and (iv) computational metrics that uses spatio-temporal data to determine some collective tendencies.

3.2.1 Traditional Notational Analysis

Traditional notational analysis based on individual variables of analysis and specific categories of actions were well researched in the last twenty years (Franks and McGarry 1996; Hughes and Franks 2005). Notational analysis has focused traditionally on team and match-play sports, studying the interactions between players and the movements and behaviors of individual team members (Clemente et al. 2013b). Therefore notational analysts have focused on general match indicators, tactical indicators and technical indicators and have contributed to our understanding of the physiological, psychological, technical and tactical demands of many sports (Hughes and Bartlett 2002).

The main objective of the notational game analysis includes optimizing feedback to the performer and coach to improve their performance (Carling et al. 2009). Thus, the information given to the coach need to be important and relevant in order to understand the reality. A well-designed system provides to the coach accurate and reliable information that is easily gathered and has an impact on subsequent practice and performance (Carling et al. 2005). Therefore performance parameters researched are one of the most important factors to provide quality to the analysis systems.

To analyze the collective performance of the teams it is important to understand and determine the relevant parameters to achieve the main goals of the observation. Thus, it is important to determine specific parameters or indicators that can provide important information to the analysts. A performance indicator is a selection, or combination, of action variables that aims to define some aspects of a performance in a given sport and, these performance indicators, should relate to successful performance or outcome (Hughes and Bartlett 2002). Therefore effective evaluation of these components requires knowledge of the contextual factors that can potentially affect the performance (Taylor et al. 2008).

Different models of the game may represent the collective tendencies to be more offensive or defensive, to act in order to attack most fast or more slowly (Clemente et al. 2013a). In fact, the model of the game and the context can influence the typology of the performance resulting in changes of the performance. Exemplifying, some studies showed that goals occurs when teams played in more direct way (i.e., goals happens with less sequence of passes) (Carling et al. 2005). This tactical approach improved the success of the some teams in the lower divisions of the English League (Hughes and Bartlett 2002). Nevertheless, the evolution of the professional football bring other styles of play with more sequences of passes and ball circulation that also revealed success (Peña and Touchette 2012).

3.2.1.1 Notational Study of Defensive Variables in Football

The main objective of defensive process is to recover the ball from the opponent (Costa et al. 2010). In a study that analyzed the recovery patterns on four matches on FIFA World Cup 1994 (Gréhaigne et al. 2001) it was verified that to be effective it is required that: (i) the defense must be in block or numerically balanced (because study shows that the recovery balls happens 90.5 % with numerical superiority and just 8.8 % with numerical equality) and; (ii) when the plays is in motion, a fixed defender is useless. Nevertheless, these kinds of rules only can be applied with a reduced number of defensive skills and in specific areas of the field.

Besides of the importance of recovery patterns to avoid suffer goals, the accuracy in defensive moments and the zones of recovery are associated with success in attack (Gréhaigne et al. 2001; Carling et al. 2005). The results in defensive moments reported that the defensive sector was the area with highest incidence of ball possession recovery (Gréhaigne et al. 2001). In the 225 actions observed, 72.5 % of the ball possession recoveries were performed at central defensive zone.

Analyzing the relationship between the zones of balls recovered and the effectiveness of the attacking process it was possible to observe that 66 % of recoveries in the defensive sector reached the forward sector (Castelo 1996). The balls recovered on defensive midfield reached in 70 % the forward sector. Finally, the balls recovered in the forward midfield sector reached in 80 % the forward sector. Thus, it was possible to suggest that one of the key factors to be efficient in the offensive process comes from the local of ball recovering (Castelo 1996).

In the study carried out in Copa America 2001 it was verified that balls recovered in the last attacking quarter resulted in 41.63 % of the shots that were made and 68.42 % of shots came from balls recovered in offensive areas (Hughes and Churchill 2005). In a older study, it was observed that 50 % of all goals came from balls recovered in the final offensive quarter (Reep and Benjamin 1968).

In a different analysis it was found that majority of goals scored in dynamic situations (i.e., open play) resulted from the balls recovered in defensive sector (Grant et al. 1999). In other study carried out in FIFA World Cup 2002, it was found that 64 % of goals emerged from balls recovered in defensive half (Carling et al. 2005).

The relationship between balls recovered and the attacking building was also recently analyzed and it was found that 51.6 % of the attacks began in the defensive half and 45.5 % on the middle half (Tenga et al. 2010). In the same study, it was found that the majority of goals resulted from the middle half.

3.2.1.2 Notational Study of Attacking Variables in Football

The goals scored are the most important variable in traditional notational analysis (Lago and Martín 2007). The notational evidences have suggesting that goals emerged with more regularity by the end of matches (Jinshan et al. 1993; Hughes and Churchill 2005). In other hand, the majority of goals suffered occurs in the last 15 min of the matches (Carling et al. 2005). These results suggest that the majority of goals resulted from a last pressing and adventurous offensive strategies to score (Carling et al. 2005).

In a different analysis it was found that shots from 30 m or more have a scoring rate of nearly 0 %, while shots from within 16.5 and 5.5 m have a scoring rate of 10 and 15 %, respectively (Dufour 1993). In FIFA World Cup 2002 the majority of the goals (37 %) were scored from inside of the penalty area, specifically the area between the edge of the 5.5 m and the penalty spot (Carling et al. 2005). Between line of goal and 5.5 m were scored 29 % of the goals. Finally, between 11 and 16.5 m it was scored 18 % of goals. For that reason, the proximity to the goal statistically increases the potential to score (Tenga et al. 2010).

Different styles of play may also influence the scoring methods. Some studies showed that direct play is the most efficient method to score (Reep et al. 1971; Carling et al. 2005). Reep et al. (1971), analyzed that 80 % of the goals scored resulted from a sequence of three passes or less. The same evidence it was found on studies that analyzed FIFA World Cup 1998 and 2002 (Carling et al. 2005). In these competitions the results showed that the majority of goals were scored by making sequences of play involving 1–4 passes. Following the previous results, the study about FIFA World Cup 2006 showed that the numbers of passes completed prior to a goal scored were 54 % after one to four passes and 29 % were after five passes or more (Acar et al. 2009).

These results may suggest that direct style after ball recovering may be better to exploit the defensive unbalance of the opponent team. By other hand, long passing sequences may provide greater opportunities rebalance the defensive process, thus minimizing the surprises and dislocation of the defense (Hughes and Franks 2005).

Considering the time expended in possession of the ball before the goal scored it was found that the majority of the goals (53 %) on FIFA World Cup 2002, were scored after periods of possession lasting between 6 and 15 s; a smaller yet significant proportion of goals were scored after ball possession lasting 0–5 s and 21–25 s (Carling et al. 2005). Similar results on FIFA World Cup 1998 revealed that a higher proportion of goals were scored after periods of possession lasting more than 21 s in 2002 (24 %) compared with 1998 (16 %). The study on FIFA World Cup 2006 showed that 72 % of the goals scored occurred at the time interval between 1 and 15 s, where 40 % were performed between 1 and 5 s with possession of the ball (Acar et al. 2009).

The relationship between possession of the ball and success have been also investigated in traditional notational analysis (Lago-Peñas and Dellal 2010). Some findings suggest that the ability to retain the possession of the ball for prolonged periods of time is associated with success in competition (Hughes and Franks 2005; Lago-Peñas and Dellal 2010; Lago and Martín 2007).

A study carried out in Spanish league revealed that the possession of the ball depends on the evolving match status (e.g., win, lose or draw), nevertheless it is found that teams have higher possession of the ball when they are losing than when they are winning or drawing (Lago and Martín 2007). Other determinants that influence the ball possession are the home matches that statistically increases the possession of the ball in 6 % in comparison with games out of home (Lago and Martín 2007). Furthermore, there are differences in ball possession depending of the team's properties and the opponent. Also in Spanish Football League it was found that possession of the ball was greater in moments of lose than in moments of won or draw (Lago 2009).

Besides the match status, it is also possible to identify the strength of the team as an important variable during possession of the ball. Some studies found that top teams retains more the ball than their opponents, thus suggesting that they prefer to control the game (Jones et al. 2004; Lago and Martín 2007; Taylor et al. 2008).

The study carried out in English Premier League found that the top three teams in the English Premier League in the 2003–2004 season (Chelsea, Manchester, and Arsenal) dominated ball possession against their opponents whether winning, losing or drawing (Bloomfield et al. 2005). For that reason, the strategies for possession of the ball can be influenced by match variables and play styles, either independently or interactively (Lago and Martín 2007; Lago 2009; Lago-Peñas and Dellal 2010).

3.2.2 Semi-computational Approaches: Observational Methodology

The semi-computational system depends on a human operator who controls and records the necessary data. The system helps the human operator to collect, store and treat the data. Nevertheless, the human operator has a preponderant influence during all the process. It is possible to identify analysis that comes from semi-automated systems: (i) tactical performance of players; (ii) network analysis of teammates; and (iii) t-patterns of collective interaction.

One example of tactical analysis using the semi-automated system is the SoccerEye (Barreira et al. 2013, 2014). This system is a software tool for observing, recording and exporting motion data to multiple formats (Barreira et al. 2014). Such a system provides two different recording designs (Barreira et al. 2013): (i) restricted recording (the observer is able to select the active categories such as the situational variables and the behavioural, spatial and interactional events); and (ii) open recording (the observed defines the observational categories of interest). In terms of this method, it is interesting to inspect the tactical behaviour of football teams. Nevertheless, each category has specific criteria that must be known by the observer. Besides background requirement about the knowledge of observational methodology, the recording process occurs during all match visualisation consuming much work and time.

The other semi-automated system is the network analysis (Lusher et al. 2010). Usually, the network analysis is performed by recording the interaction between teammates throughout the match (Bourbousson et al. 2010). Such interaction can be defined by a linkage variable such as the passes between teammates (Bourbousson et al. 2010), the displacements (Passos et al. 2011) or the defensive coverage (Duarte et al. 2012). The majority of works which performed network analysis used paper-and-pencil notation or even a general semi-automated system such as Amisco©. Nevertheless, in all cases, it is necessary to build an adjacency matrix where the entries represent the linkage between teammates (Clemente et al. 2015a, b, c, d). By using such a matrix and giving to it a weight (Horvath 2011), it is possible to compute a set of network metrics that characterise the relationship within the team, as well as identify the prominent players in the attacking or defensive strategy. The potential and practical applications of network analysis are too great. It is possible to understand the inter-relationship dynamics in the team, identify the players that highly contribute to the defensive or attacking process or even to characterise some clusters in the team. Such information can be very useful for coaches during the training process or even in matches. Nevertheless, the user's building of the matrix of cooperation requires much time and work, thus the possibilities to use in an online-fashion are too low. Thus, the potential is mainly for post-match analysis.

The last semi-automated system is the t-patterns (Magnusson 1996). Such analysis is used to assess the temporal patterns that occur during the teammates' interaction. These temporal patterns can be detected by using a specific algorithm of the THEME software (Magnusson 2000). To inspect the t-patterns of football teams a dedicated software was built called SOF-CODER (Jonsson et al. 2006). This software consists of manual coding. The coding depends on the user's work, which occurs during the digital match visualisation. Thus, the SOF-CODER is a tool that is used both for observation and recording processes (Jonsson et al. 2006). The t-patterns are computed using the THEME software. The outcomes provided by the t-patterns recognition are actually interesting, thus giving to the observer an estimation of teammates' patterns of interaction. This information can be useful to detect some patterns of play in a team, to identify its properties of attacking build. Once again, the weakness is the time spent during the data collecting and processing. In fact, the human operator is required to perform the data collection, making it too difficult to carry out such analysis during matches (online).

In sum, in all semi-automated systems previously presented, the recording process still needs the user's selection. It seems obvious that these systems are better than the traditional paper-and-pencil process. Moreover, their practical applications are huge; they can be used in any situation, only requiring the software and the video-recording of a match. Thus, any coach or analyst can use such a system to observe a team's behaviour even an amateur team. Moreover, the information collected is specific to a football game in accordance with the specific principles of play. Nevertheless, the great amount of time spent on observation is an advantage. The online observation (during a game) is also very difficult due to the complexity of data recording and process. Thus, despite the many advantages and practical applications of these semi-automated systems for football, match analysis still awaits quick, automatic and user-friendly systems.

3.2.3 Computational Approaches: Spatio-Temporal and Tactical Metrics

Novel estimation, detection and identification techniques have been recently applied on sports, providing the Cartesian positional information of players over time (Clemente et al. 2013b, 2015a, b, c, d). This information has been seen as vital within sports science's literature, so as to propose new computational tactical metrics that may allow to inspect the spatio-temporal relationship between teammates (Bartlett et al. 2012). Such technological approaches can improve the understanding of the collective match, providing to coaches and analysts a real-time augmented perception of the game (Clemente et al. 2014c).

It is possible to observe that all these new technological metrics need to be understandable by a great range of coaches and analysts. In fact, the user-friendly system must be the essence of such metrics. Moreover, the opportunity to collect simple and pertinent information should be taken into account for the system to be generalised by all the football community. There must be a threshold between the complexity of such metrics and the applicability of information for coaches and analysts.

These metrics have a valuable strength in comparison with the semi-automated systems. The use of an integrated system that is totally autonomous from the data retrieval to the data processing allows using such analyses during official matches or even in daily training sessions without a great effort on the part of a human operator. Nevertheless, one main issue can be discussed. In fact, the system depends on an automatic tracking method that is now too expensive for amateur or even some professional teams. Thus, the tracking method must be prioritised in an integrated system.

One of the first metrics proposed was de centroid (Frencken and Lemmink 2008) and then the wCentroid (Frencken et al. 2011). This metric represents the centre-of-mass of the team at a given instant based on the proximity of each player with the ball position. Using such information the coach can identify the strong point of the team at a given moment. This point can be useful to identify the global position of a team during the match. Some studies showed that when the Centroid of a team with possession of the ball overcomes the defensive Centroid, the possibilities to shoot or score increase (Frencken et al. 2011). It was further noted importantly that the wCentroid decreases their position from the first to the second half of the match (Clemente et al. 2013b). The authors of this study suggested that fatigue can be one of the causes for such variation (Clemente et al. 2013b). At any rate, the most important information provided by wCentroid is identification of the central point of the team, thus representing its global position. In comparison with the opponent's wCentroid, it is possible to identify the in-phase or anti-phase pattern with the variation of the opponent's wCentroid. Another important pattern that can be detected is the tendency to act on a specific side of the football field.

Using the wCentroid, it is possible to extend such an approach to assess the players' dispersion over their centre. Thus, the wStretch Index it was proposed to measure the dispersion of players in both attacking and defensive moments (Clemente et al. 2013b). The dispersion of teammates is a very important indicator of collective organisation. In fact, during the defensive moments a concentration behaviour is required where players are closer to each other, trying to reduce the possibilities of penetration. On the other hand, during the attacking moments, the team disperse over the field trying to avoid the opponent's marking and to create opportunities to penetrate and score. Such concentration and dispersion is usually called expansion-contraction behavior (Moura et al. 2012). Thus, using the wStretch Index, it is possible to assess the dispersion level of a team and to identify quickly whether it is a regular pattern in attacking or defensive moments. In fact, a development of this metric is to introduce some warning notification for great dispersions during defensive moments or for small dispersions during attacking moments. It is also possible to identify some patterns and characterise the team's dispersion throughout the match.

The concepts of Surface Area and Effective Area of Play were also introduced and discussed in the last few years (Okihara et al. 2004; Clemente et al. 2013b; Frencken et al. 2011). The surface area generates the minimum number of triangulations between all the teammates, thus drawing a polygon that defines the covered area of a team (Gréhaigne 1992). Using such metric, it is possible to have a similar measure to the wStretch Index. Nevertheless, in this case, the measure is the area covered by all triangulations. However, such information is not enough for coaches mainly to use during matches. In fact, some indicators of efficacy lack surface area. Thus, a new concept was proposed to estimates the triangulations between teammates and also identifies the overlapping triangulations between both teams (Clemente et al. 2013b). To assess the effective triangulations in the over-lapping situation, it was proposed a maximal perimeter of defensive triangulations. Thus, the overlapped triangulations with more than 36 m of perimeter in defensive moments are considered non-effective because the pressure is too much reduced. Obviously, this one player can recover the ball, but the criteria were generated taking into account the defensive coverage that assumes a greater proximity of teammates to the player in defensive delay process (Costa et al. 2009). Using such metric, it is possible to have qualitative criteria mainly during defensive moments. Moreover, the variable of 36 m perimeter can be adjusted to the coach's require-ment. Such information allows quick identification, using the graphical interface, of the effective triangulations and during a match reorganisation of the collective synchronisation and optimisation of their tactical behavior (Clemente et al. 2013b).

If the wCentroid, wStretch Index, Surface Area and Effective Area of Play can be considered a spatio-temporal metric that evaluates the teammates' synchroni-sation, the territorial domain was proposed as identifying the numerical relationship between opponent teams within a specific region of the field. This metric was first applied in a professional football match (Vilar et al. 2013). The potential of this metric is very interesting as it mainly identifies the strategic distribution of a team and characterises the strength and weakness regions of the field of one team. In this regard, an algorithm was proposed for the seven-a-side game that is used in youth football (Clemente et al. 2015a, b, c, d). In fact, the applications for the coaches of younger teams is fundamental, using this metric to identify whether the players have a rational distribution or whether they play in an agglomerate way. Moreover, such metric can identify how the team acts to reduce the numerical disadvantage in vital regions such as the central defensive area or central midfield.

Besides the previously discussed metrics that can be used in football or even in other invasion team sports (e.g., basketball, futsal, handball), it has been proposed a set of metrics that measure the specific football principles of play. A set of criteria using players' location would turn the semi-automated systems into an automatic one. Taking as a reference the semi-automated system (FUT-SAT) to assess the principles of play (Costa et al. 2010), it has been proposed a full automated attacking metrics inspired in some criteria used by the manual notation (Clemente et al. 2014a, b). These metrics were also developed based on the information about players' location during the match. Using the criteria to identify the efficacy of the team to accomplish each principle, it was possible to create seven ratios that change during each attacking unit (the instant between the first pass to the loss of the ball).

In the last few years it were proposed the ratios of penetration, attacking cov-erage, coverage in support, coverage in vigilance, depth mobility, width and length and attacking unit (Clemente et al. 2014a, b). All of these metrics assess the efficacy of the team to accomplish during each attacking attempt the fundamental football tactical principles of play. The main findings of these metrics showed that the majority of attacking coverage is performed in vigilance and not in support, thus suggesting that the analysed team opts for a more direct style of play. This suggestion can be enhanced by the great ratio of depth mobility that means a great longitudinal dispersion of the forward players. Another interesting finding was the great ratio of unit principle of play in attacking moments, thus suggesting a notable synchronisation of all teammates with the movement of ball and forward players. These tactical metrics make it possible to define the accomplishment of tactical principles and to identify some specific properties of a team. Obviously, these relative ratios can be complementary information and from a graphical point-a-view are ineffective. In spite of the fragilities during the match, these metrics can provide important information to classify the team's properties after the games.

The last proposed tactical metric in the literature was the defensive play area and the sectorial lines (Clemente et al. 2014a, b, 2015a, b, c, d). These metrics were proposed based on the method that estimates the momentary tactical mission of each player in the game. The momentary tactical mission was developed based on the concept that the variability of players' motion determines that their mission is different from the defensive for attacking instants. Such variability implies that their tactical participation can be different from instant-to-instant, i.e., a central defender can act as a striker for a few moments. Thus, it was proposed a method to estimate their momentary tactical mission based on the players' location. This made it possible to define the defensive regions of pressing. These regions were based on manual notation (Seabra 2010). Thus, different regions of defensive pressing were classified based on the triangulations performed by the teammates and on each momentary tactical mission performed by each player that constitutes the triangulation. It was found that the highest defensive area of pressing was covered in the second half of the midfield region. It was also found that the second highest region of defensive pressing was performed in the first half of the midfield region. Thus, the greatest amount of defensive pressing was performed midfield, a crucial element to be covered in defensive moments. This metric can be an interesting solution to use in an online-fashion, quickly identifying each different region of pressing with different colours. Therefore, their graphical interface with the coach or analyst can be better than the final outcomes.

Besides the defensive play area, it was also proposed the sectorial lines. Such metric uses the information retrieved by the momentary tactical mission to identify the line of play of a given sector. This metric was based on the concept of axes (Gréhaigne 1992; Lemoine et al. 2005), which was used to define the synchronisation of defensive and attacking axes of a set of players involved in the play in the moments before the shot at goal. The sectorial lines extended the concept of axes to all the team and during the match using an automatic method (Clemente et al. 2014a, b). A correlation test was carried out to inspect the synchronisation between the angles of the different sectorial lines (defensive, midfield and forward). It was found a small correlation between the angles of sectorial lines, thus suggesting that each region acts with a relative independence from the other sectors. It was also found that the angle of sectorial lines is more neutral during defensive moments and increases during the attacking moments due to the specific tactical mission performed by the wings' exploitation. Similarly to wCentroid, wStretch Index, Surface Area, Effective Area of Play, Territorial Domain and Defensive Play Area, the Sectorial Lines can be graphically used by coaches during a match to enhance the perception about their team behavior and even about the opposing team.

Despite of the possibilities provided by computational and automatic systems, the high-price and the non-possibility to use in any competitive level reduces their practical applications. For that reason, a mixed approach between traditional analysis and computational analysis can be the most common process to improve the possibilities of match analysis for coaches. In this context, social network analysis reveals a set of possibilities to identify the general properties of graphs and also measure the centrality levels of players. For that reason, in the next section will be discussed how it is possible to generalize the social network analysis to the majority of team sports community.

3.3 Building a Low-Cost Instrument of Observation for Network Analysis

The low-cost solutions for match analysis using social network analysis (SNA) must follow some requirements to ensure the reliability of the data. In the specific case of team sports, the variability of actions and the variance in the contexts may lead to subjectivity (Franks and McGarry 1996). Such subjectivity will compromise the data collecting and the interpretation of the results. For that reason, a test-retest to intra- and inter-observers must be carried out to avoid errors and differences between observers. Generally, a test-retest with 20 % of the sample will ensure the reliability of the data. In some cases, the minimum of 10 % of the full data is enough to test the reliability of procedures (Tabachnick and Fidell 2007). Another alternative it is follows a protocol of teaching with 15–30 h in order to consolidate the criteria of analysis. After such program, the same sample of test must be analyzed in two occasions with 25-days interval (Robinson and O'Donoghue 2007). The interval of two or three weeks will minimize the observer's familiarity with the task (Altman 1991).

After data collection, the reliability test can be tested using the Kappa of Cohen coefficient (Robinson and O'Donoghue 2007). In the case of more than one observer, the test-retest must be analyzed for intra-observer (the reliability for the same observer in two occasions) and for inter-observers (the reliability between different observers in two occasions). Values greatest than 0.61 in Kappa of Cohen coefficient will ensure the reliability of the data, thus indicating a conventional level of acceptance (Landis and Koch 1977). In other hand, if a lower value is achieved during the test, another round of observers' training must be performed.

3.3.1   Defining the Variables

The most important criteria to develop a match analysis system are to define the variables. The variables will constrain the procedures to collect the data and the observational process. Moreover, in the case of SNA it is always need a linkage criterion to connect two nodes. This criterion will depend from the requirements of coach of the scientific analysis.

Usually, in the cases of no software the systems must be carried out in three categories (Hughes and Franks 2004): (i) scatter diagrams; (ii) frequency tables; and (iii) sequential systems. Following the authors' idea, scatter diagrams are usually simple and are most often used to gather data in-event and enable immediate feedback for the coach and athlete (Hughes and Franks 2004). Therefore, a scatter diagram involves drawing or record a schematic representation of the playing surface of the sport, and then notating on this the actions of interest, at the position at which they took place (Hughes and Franks 2004).

In other hand, frequency tables are a commonly used form of data gathering that enables quick, simple analyses of performance of athletes and teams. Finally, the record of sequence in which events occur enables the analyst to go to far greater depths in interpreting a performance (Hughes and Franks 2004). This sequential data system will help to identify the patterns of actions based on the notion of temporal sequences. Following the three categories can do data recording by using SNA. Despite the complexity of analysis, the categories will provide different perspectives for coaches and also for data interpretation in scientific point-of-view. For that reason, let us consider an example of basketball analysis for the linkage criteria of successful passes performed between point guard (PG1) to the remaining teammates (P2, P3, P4 and P5) in a specific context with three sequences of attacking Fig. 3.1.

As possible to observe in the above Fig. 3.1, the traditional notation allowed to identity the zones of passes made by the player and the tendencies of interaction in attacking moments. Nevertheless, the SNA requires adjacency matrices to make possible compute the following metrics. For that reason, the following Table 3.1 will show the initial two adjacency matrices of the passing sequences.

This is one of the small possibilities of analysis that can be made using notational analysis. Nevertheless, another issues must be managed to choose the variables. In the specific case of SNA the linkage criterion must be exact and with a small range of subjectivity. One of the most common linkage indicator that have been used by SNA in team sports are the passes performed for the teammates (Peña and Touchette 2012; Cotta et al. 2013; Duch et al. 2010; Clemente et al. 2015a, b, c, d). In one case it was analyzed the direction of passes between sectors (Malta and Travassos 2014). Maybe the main reason to the same kind of analysis can be the high level of subjectivity in the remaining possibilities. Let us provide some examples for defensive actions and attacking actions in Fig. 3.2.

As possible to observe in Fig. 3.2, there are other linkage indicators that can be used to define as variables in the SNA. Nevertheless, these cases are also more difficult to analyze than passes made. In the example 1 the aim is to analyze the

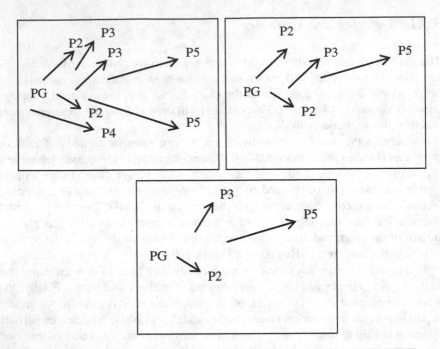

	Sequence 1	Sequence 2	Sequence 3	Total
P2	2	2	1	5
P3	2	1	1	4
P4	1	0	0	1
P5	2	1	1	4

Fig. 3.1 Sequences of play and notation of passes made from PG1 to the remaining teammates

Table 3.1 Corresponding initial two-adjacency matrices for the interaction from PG1 to the remaining teammates

	PG1	P2	P3	P4	P5	PG1	P2	P3	P4	P5
PG1	0	2	2	1	0	0	2	1	0	1
P2	0	0	0	0	0	0	0	0	0	0
P3	0	0	0	0	0	0	0	0	0	0
P4	0	0	0	0	0	0	0	0	0	0
P5	0	0	0	0	0	0	0	0	0	0

passes not completed. It is possible to observe an attacking sequence with two attackers providing lines of passes to the player with possession of the ball. In the sequence of play, the opponent intercepted the ball but the link of not completed passes is obvious? It is evident the idea of pass made by the player with possession of the ball? The trajectory and movement of the ball is a helpful contribution to

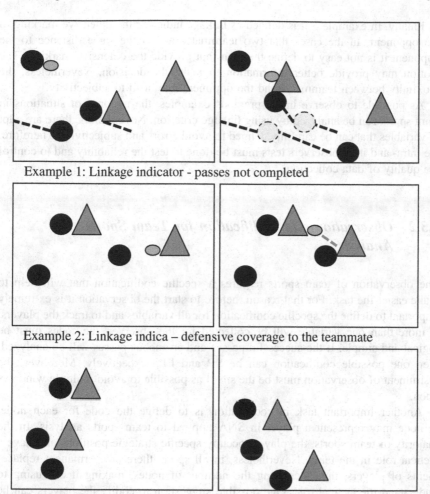

Example 1: Linkage indicator - passes not completed

Example 2: Linkage indica – defensive coverage to the teammate

Example 3: Linkage indicator – defensive marking to the opponent

Fig. 3.2 Examples of linkage indicators that can be analyzed in team sports in a point-of-view of SNA

analyze the link and the idea of pass. Nevertheless, in some cases the proximity between teammates do not make easy to identify if player A made the erroneous pass to B or to C.

In the example 2 the linkage indicator was defined as the defensive coverage made by a teammate to the other that are marking the opponent with possession of the ball. In the case, it is not easy to identify the player that provides the defensive coverage to the teammate, because both teammates may help in the situation. Nevertheless, in the sequence of play it is only one hypothesis that the backward player provides the defensive coverage to the player closest to the ball and the opponent that will receive the ball.

Finally, in example 3 it is defined as linkage indicator the defensive marking to the opponent. In the cases that two teammates are in the same distance to the opponent it is not easy to define the man that provide the defensive marking. The motion may provide better information to make the decision. Nevertheless, the proximity between teammates and the opponent may lead to subjectivity.

As possible to observe in the previous examples, the majority of situations in team sports can be analyzed by using linkage criterion. Nevertheless, there are a lot of variables that can be carefully defined to avoid errors and subjectivity. Therefore, the intra- and inter-observers tests must be done to test the reliability and to control the quality of data collection.

3.3.2   Observation and Codification for Team Sports Analysis

The observation of team sports requires a specific codification that will help to make easier the task. For that reason, before to start the observation it is extremely important to define the specific codification for all variables and to track the players. If more than one variable will be collected for the observer, the codes must be logical and simple. If the successful passes and erroneous passes will be analyzed, then one possible codification can be SP and EP, respectively. Moreover, the instrument of observation must be the small as possible to avoid multiple windows open.

Another important task in codification is to define the code for each node. A node may represent a player in SNA applied to team sports analysis. In the majority of team sports, the players occupy specific strategic positions and have a tactical role in the team. Nevertheless, in all sports there are permanent replacements of players, thus increasing the number of nodes, making it confusing to understand the tactical role. One possible suggestion to codify the players can be attributed a specific number per each player, thus when a new player enters the match, a new node is added to the graph. However, this solution may represent an issue because in an example of basketball, the team may lead to ten nodes not representing the tactical context. To avoid this issue it is possible codify players by their strategic position on the field. This possibility requires an accurate observation to identify the tactical position of each player. Moreover, the observer should be alert to occasional changes in strategic distribution of the team that occurs in the match. This possibility ensures that only the tactical roles matters independently of the player in question. For that reason, this possibility may be very useful to identify a team's style of play, patterns of interactions and also some specific tactical characteristics.

These possible solutions to code the players have advantages and disadvantages. In the case of codification by tactical role will be easier to identify the position and to increase the accuracy of data collection. Moreover, this will help to increase the reliability of centrality metrics. Nevertheless, the tactical position does not allow identifying which player influenced a specific pattern of interaction. Only with a temporal analysis will be possible to solve this issue. Moreover, the information about a specific player will be missing in the tactical role assumed by other players during the replacement process that occurs in match. By other hand, when the team is observed in a strategic and tactical point-of-view, it is possible to identify some properties of the team that are not exclusively dependent from the individual player. Thus, the codification by tactical role will be benefit to identify the patterns of play and the strategic plans of the team. Nevertheless, it is possible to suggest that the analysis by tactical role is hardest to observe in practice because teams may change their strategy and even players change their strategies permanently, thus the observer must be much trained for this. A consistent test-retest process may be carried out before this observation.

The alternative to observe the players by their name or code will miss the tactical information because more nodes will appear in the graph. As described before, in the basketball analysis may appear ten nodes. Thus, the tactical information is missing in this situation. Therefore, in order to overcome this issue will be necessary a specific temporal analysis every time that one player replace the other. This alternative will help to ensure that only the players in match will be analyzed in the team's structure. The following Fig. 3.3 will represent both situations.

The case of example A follows the codification based on tactical position. In the weighted graph it is possible to observe the greatest links by the size of arrows and also the overall connectivity. In the point-of-view of coach is easier to analyze the first example. Moreover, the centrality metrics and macro analysis will represent the participation and interactions with specific values. In other hand, in the example B it is possible to observe the weighted graph of all players that participated in the match. This graph is hardest to visualize, nevertheless it is possible to identify the centrality property of a given player.

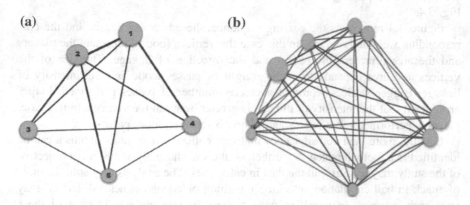

Fig. 3.3 Example **a** Graph of basketball analysis using a codification based on tactical position; example **b** graph of all basketball players that participated in the match

3.3.3   Collecting and Processing the Data

It is required in SNA an observation that results in adjacency matrix, as previously described. Using such matrix will be possible to identify the arcs as the linkage indicator that may assume an action such as defensive coverage, defensive marking or ball recovery (from one player to the opponent). As visualized above, the pass is the easiest indicator to observe in SNA of team ports. In any team sport the pass is the most common individual action performed between teammates, which keeps possession of the ball while moving it forward to score. For that reason, following we will provide an example based on passes performed.

It was defined as our criteria to split the data per each unit of attack (passing sequence without interception of the ball). During a match it is possible to observe between 150 to more than 300 passes made by the same team (Carling et al. 2005), for that reason the split by units of attack will help to identify the temporal patterns. An attacking unit can be classified by noting the moment of ball recovery, followed by a set of passes performed without losing the ball (Passos et al. 2011; Clemente et al. 2015a, b, c, d). Following, if the ball is lost this attacking unit ends and we generated a specific adjacency matrix of the attacking unit (Clemente et al. 2015a, b, c, d). To generate a adjacency matrix per unit of attack, it was considered that a sequence of passes should integrate more than three successful passes (Malta and Travassos 2014). It was used a codification of one (1) to notate a pass from one player to another and zero (0) for no passes, as described in previous works (Passos et al. 2011; Clemente et al. 2015a, b, c, d). In the cases of more than 1 pass was performed in the same direction (from player 1 to player 2) the code equalled the number of passes in the same attacking unit (Clemente et al. 2015a, b, c, d). In team sports, the weighted digraphs are the best approach to measure the passing sequences. The passes represents a direction (player C to player D is different than D to C) and that the number of passes per direction represents a weight. Following the example of passing sequence, let us provide an illustrative representation in Fig. 3.4.

Figure 3.4 represents the passing sequence, the adjacency matrix and the corresponding weighted digraph. In this case the vertices (nodes) represent the players and the arcs represent the passes and the direction of linkage. The size of the vertices and arcs depends from the weight of passes made and the intensity of linkage. Bigger vertices represent a greater number of passes performed. Larger arcs meant that the intensity of linkage is greater between the vertices. In this case, player 1 performed more passes and player 3 received more passes.

Usually, more than 100 adjacency matrices collected from units of attack can be identified in a football match. Nevertheless, the specific analysis system or objective of the study may lead to split the data in categories. The analysis may split the units of attack in ball circulation, attacking transition or counter-attack. Moreover, may split each category in attacking with success or non-success. Despite of these categories, a temporal analysis by periods of time may also be interesting in order to identify the patterns of interaction in different contexts. Nevertheless, if only a

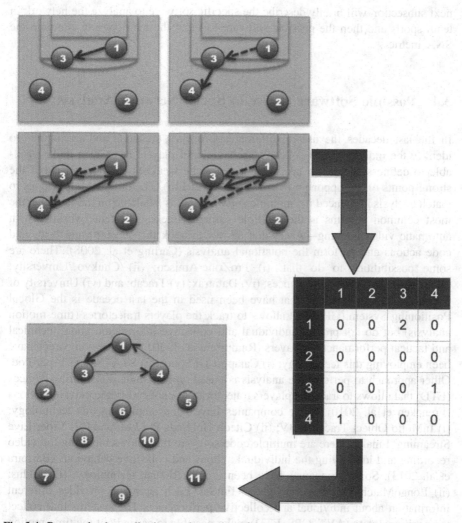

Fig. 3.4 Process the data collecting: **a** observation of the real passing sequence; **b** converting the observation for the corresponding adjacency matrix; **c** converting for the weighted digraph

global analysis is required, a representation per each half or per the complete match can be made. Briefly, the adjacency matrices must be added to a final adjacency matrix once analysts decide which the kind of analysis they will perform.

As possible to verify, all these steps were made without computation of SNA metrics. The observation can be more or less complex and depends from the objectives of the observer. This section described a low-cost process to collect data without software of analysis. In the end of observation will be possible to insert the matrices in specific SNA software. By other hand, there is specific software to track the interactions in team sports that will be following presented. For that reason, the next subsection will briefly describe the specific software to analyze the network in team sports and then the possible software to input the matrices and compute the SNA metrics.

3.4   Possible Software to Use in Social Network Analysis

In the last decades the use of different approaches, methods and techniques to identify the individual and collective behaviour of players have been make available to define strategies to take advantage of the weakness points and avoid the strong points of the opponent (Sarmento et al. 2014b). The different approaches to match analysis influenced the multiple software that is available on market. One the most common systems is the multiple cameras tracking system (working with automatic video tracking—AVT) that allows to track the players trajectories and code actions and perform the notational analysis (Carling et al. 2008). There are some possibilities to do that: (i) ProZone-Amisco; (ii) Chukyo University; (iii) Hiroshima College of Sciences; (iv) Datatrax; (v) Tracab; and (vi) University of Campinas. Another system that have been used in the last decade is the Global Positioning System (GPS) that allows to track the players trajectories (time-motion analysis) but do not provide individual and collective information about technical and tactical performance of players (Rampinini et al. 2014). Some companies have been employing this technology: (i) Catapult-GPSport; (ii) STATSports Viper Pod. Other approach to perform the analysis is based on electronic transmitting devices (ETD) that allows to track the players trajectories but not codify the players' actions (Frencken et al. 2010). Some companies have been employing this technology: (i) InMotio Object Tracking BV; (ii) Citech Holdings Pty Ltd; and (iii) Viper Live Streaming. Finally, there are multiple code software that allows inputting the video recording and identifying the individual actions and collective behaviour (Barreira et al. 2013). Some companies have been employed such technology: (i) Noldus; (ii) LongoMatch; (iii) SoccerEye; (iv) FutSat. Each approach provides different information about individual and collective performance. The systems that collect the positional data (AVT, GPS, ETD) make possible to carried out the time-motion analysis (Couceiro et al. 2013). This analysis provides information about the distance covered, the speed, the time spent in a specific intensity of running speed, the distance covered in specific intensity of running speed, acceleration, deceleration and change in directions (Carling et al. 2005). Such information give to coaches and analysts the global picture about the physical demands of each tactical position, as well as the specific fitness level of players during the match and training sessions (Clemente et al. 2015a, b, c, d). By other had, the systems based on video cameras allows codifying the individual actions (notational analysis) and collective behaviour (using qualitative methods) (Sarmento et al. 2014a; Barreira et al. 2014). Such notational process is based on human operator work in online (during) or offline (after match) fashion (Clemente et al. 2015a, b, c, d).

For the case of SNA applied to sports, the semi-computational methods based on human observer are the most adequate. Nevertheless, only one specific software it was developed to SNA in sport, so far. For that reason, following will be introduced the Performance Analysis Tool.

3.4.1   Performance Analysis Tool (PATO)

The Performance Analysis Tool (PATO) is a new scientific software to code the interactions made in team sports (Clemente et al. 2015a, b, c, d). For a deep revision about PATO's software please consult the full article "Performance Analysis Tool for network analysis on team sports: A case study of FIFA Soccer World Cup 2014".

Briefly, in this software it is possible to choose the number of players that observer wants to analyze. This will help the observer to code any kind of games with more than three players. After that, the system will show by default a layout with two windows that represents the two teams in match. To move from one to the other is only necessary to press the top bar with finger (in app for tablets) or with courser.

The PATO's layout have three types of specific events that user may chose in the moment of analysis (Fig. 3.5).

Let us consider the analysis for the passing sequences. After to press the players that participated in the attack, the user chooses the type of event (as example: shots, special, goal) and then "add". The sequence of playing will be added in the adjacency matrix. If the user wants to analyze between periods of 5 min, in the end of this time only need to press the bottom "save" and then continue the analysis for the remaining periods. The 5-minutes adjacency matrix will be automatically exported in TXT format that will allow using in the majority of regular SNA software. After the TXT exporting the user has two possibilities: (1) continue the observation without removing the previous adjacency matrix, thus making it possible to sum all the matrices from the beginning or (2) reset the system in order to build the adjacency matrices from 0 (Clemente et al. 2015a, b, c, d).

This simple and user-friendly solution will allow using the TXT in regular SNA software to process the analysis. This will allow to user make the decision about the analysis that want to make and the specific software that fits in the aim of the analysis. Following, a brief overview about the SNA software will be made.

3.4.2   A Brief Overview of the Available SNA Software

With an adjacency matrix resulted from the observation will be possible to start the data computation and analysis. This step may be performed by the observer or researcher in a programming tool or software development tool or can use specific

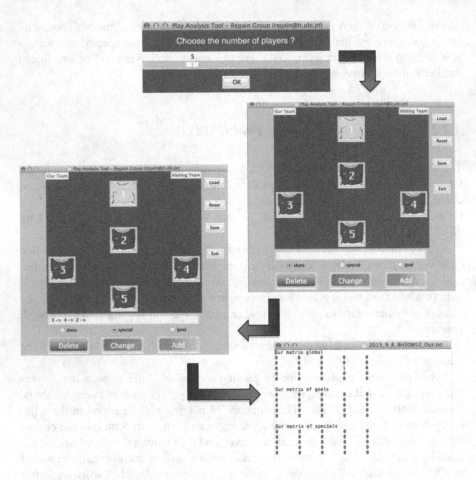

Fig. 3.5 Sequence of coding until the adjacency matrix in PATO's software

software for SNA. Trying to make easier a quick overview, let us following present some possible software for SNA analysis (Table 3.2).

As possible to observe in the above table, there are multiple alternatives to perform the SNA. Each software has pros and cons, for that reason the user should consider to try each software per each contextual requirement. It is also important to highlight that some of these software are not free and will be necessary to study the best choose before buying the license.

In the software will be possible to analyze the general properties of graph in a macro-analysis. In a meso-analysis will be possible to identify the specific types of interactions. Finally, in a micro-analysis will be possible to identify the centrality levels of a player and the individual contribution for the network. All these properties and metrics will be further described in the next chapters.

Table 3.2 Summary of the main software for SNA

Software	Functionalities	Operating system	Observations
EgoNet Source: http://sourceforge.net/projects/egonet/	Is a program for the collection and analysis of egocentric network data	Windows, Mac and Linux	Create the questionnaire, collect data, and provide general global network measures and data matrixes that can be used for further analysis by other software
Commetrix Source: http://www.commetrix.net	Software framework for dynamic network visualization and analysis	Windows	Extracting virtual communities in electronic communication networks; Analyzing dynamic network change, properties, lifecycles, and structures Creating rich expert network maps or recommendation systems from communication logs or other network data sources
Gephi Source: https://gephi.github.io	Allows an interactive visualization and exploration platform for all kinds of networks and complex systems, dynamic and hierarchical graphs	Windows, Mac and Linux	It is possible a representation, manipulate the structures, shapes and colors to reveal hidden properties. It uses a 3D render engine to display large networks in real-time and to speed up the exploration
NetMiner Source: http://www.netminer.com/main/main-read.do	Software tool for exploratory analysis and visualization of large network data	Windows	Analysis of large networks, comprehensive network measures and models, both exploratory and confirmatory analysis, interactive visual analytics, what-if network analysis, built-in statistical procedures and charts
NetworKit Source: https://networkit.iti.kit.edu	NetworKit is a growing open-source toolkit for high-performance network analysis	Mac and Linux	Its aim is to provide tools for the analysis of large networks in the size range from thousands to billions of edges
Polinode Source: https://www.polinode.com	It is a flexible tool that helps cut through complexity. At its core is the ability to map, visualize and analyze network data	Windows, Mac and Linux	Calculate measures of centrality directly in your browser Automatically detect communities with our community detection algorithm

(continued)

Table 3.2 (continued)

Software	Functionalities	Operating system	Observations
ORA Source: http://www.casos.cs.cmu.edu/projects/ora/software.php	It is a dynamic meta-network assessment and analysis tool.	Windows	It contains hundreds of social network, dynamic network metrics, trail metrics, procedures for grouping nodes, identifying local patterns, comparing and contrasting networks, groups, and individuals from a dynamic meta-network perspective
Social Network Visualizer Source: http://socnetv.sourceforge.net	It is a cross-platform, user-friendly application for the analysis and visualization of Social Networks in the form of mathematical graphs	Windows, Mac and Linux	Enables to edit the social network data through point-and-click, analyze their social and mathematical properties, produce reports for these properties and apply visualization layouts for relevant presentation of each network
UCINET Source: https://sites.google.com/site/ucinetsoftware/home	It is a software package for the analysis of social network data	Windows	UCINET is a comprehensive social network analysis tool which comes with the integrated visualization tool Netdraw (available free separately). It has a comprehensive help system

Key points

- Team sports are complex and dynamic systems that depends from permanent interactions that can be analyzed by using social network analysis;
- Match analysis provides relevant information to optimize the training sessions and to improve the strategy for the match. For that reason, the observational process it is an requirement of coach to help him to make decisions;
- The analysis by using network procedures can be a low-cost and user-friendly possibility to generalize the semi-computational analysis to all team sports in any competitive level;
- The observer may use an observational system made by him. To do that, it is only required a specific definition of variables and linkage indicators that ensures the reliability of the data collected;

- Social network analysis in team sports it is commonly used to analyze the passes performed between teammates. Nevertheless, other linkage indicators such as defensive marking, defensive coverage or attacking coverage can be also analyzed.
- The codification of players can be made by considering the tactical position or using the specific code of a player. The first alternative provides a better analyzing about the patterns of play and collective organization of team. In other hand, the second alternative provides information about the specific contribution of a player for the team's network;
- The adjacency matrices that come from observational process can be used in different SNA software to process macro-, meso- and microanalysis.

References

Acar, M. F., et al. (2009). Analysis of goals scored in the 2006 World Cup. In T. Reilly & A. F. Korkusuz (Eds.), *Science and football VI* (pp. 235–242).

Altman, D. G. (1991). *Practical statistics for medical research*. London, UK: Chapman & Hall.

Barreira, D., et al. (2014). Ball recovery patterns as a performance indicator in elite soccer. *Proceedings of the Institution of Mechanical Engineers, Part P: Journal of Sports Engineering and Technology, 228*(1), 61–72.

Barreira, D., et al. (2013). SoccerEye: a software solution to observe and record behaviours in sport settings. *The Open Sports Science Journal, 6*, 47–55.

Bartlett, R., et al. (2012). Analysing Team coordination patterns from player movement trajectories in football: Methodological considerations. *International Journal of Performance Analysis in Sport, 12*(2), 398–424.

Bloomfield, J., et al. (2005). Temporal pattern analysis and its applicability in Soccer. In L. Anolli, et al. (Eds.), *The hidden stucture of interaction: From neurons to culture patterns* (pp. 237–251). Amsterdam, The Netherlands: IOS Press.

Bourbousson, J., et al. (2010). Team coordination in basketball: Description of the cognitive connections among teammates. *Journal of Applied Sport Psychology, 22*(2), 150–166.

Carling, C., et al. (2008). The role of motion analysis in elite soccer. *Sports Medicine, 38*(10), 839–862.

Carling, C., Reilly, T., & Williams, A. (2009). *Performance assessment for field sports*. London: Routledge.

Carling, C., Williams, A. M., & Reilly, T. (2005). *Handbook of soccer match analysis: A systematic approach to improving performance*, London & New York: Taylor & Francis Group.

Castelo, J., (1996). *Futebol a organização do jogo: como entender a organização dinâmica de uma equipa de futebol e a partir desta compreensão como melhorar o rendimento e a direcção dos jogadores e da equipa*, Jorge Castelo.

Clemente, F. M., Silva, F., Martins, F. M. L., et al. (ahead-of-print). Performance analysis tool for network analysis on team sports: A case study of FIFA Soccer World Cup 2014. *Proceedings of the Institution of Mechanical Engineers, Part P: Journal of Sports Engineering and Technology*.

Clemente, F. M., Couceiro, M. S., Martins, F. M. L., et al. (2013a). Measuring collective behaviour in football teams: Inspecting the impact of each half of the match on ball possession. *International Journal of Performance Analysis in Sport, 13*(3), 678–689.

Clemente, F. M., Couceiro, M. S., Fernando, M. L., et al. (2013b). Measuring tactical behaviour using technological metrics: Case study of a football game. *International Journal of Sports Science & Coaching, 8*(4), 723–739.

Clemente, F. M., Martins, F. M. L., et al. (2014a). Developing a football tactical metric to estimate the sectorial lines: A case study. In Murgante et al. (Eds.), *Computational science and its applications* (pp. 743–753). Berlin: Springer.

Clemente, F. M., et al. (2014). Inspecting teammates' coverage during attacking plays in a football game: A case study Inspecting teammates' coverage during attacking plays in a football game: A case study. *International Journal of Performance Analysis in Sport, 14*(2), 1–27.

Clemente, F. M., Couceiro, M. S., et al. (2014c). Practical implementation of computational tactical metrics for the football game: Towards an augmenting perception of coaches and sport analysts. In B. Murgante et al. (Eds.), *Computational science and its applications* (pp. 712–727). Berlin: Springer.

Clemente, F. M., Martins, F. M., Couceiro, M. S., et al. (2015a). Developing a tactical metric to estimate the defensive area of soccer teams: The defensive play area. *Proceedings of the Institution of Mechanical Engineers, Part P: Journal of Sports Engineering and Technology.* Available at: http://pip.sagepub.com/lookup/doi/10.1177/1754337115583198.

Clemente, F. M., Martins, F. M. L., et al. (2015b). General network analysis of national soccer teams in FIFA World Cup 2014. *International Journal of Performance Analysis in Sport, 15*(1), 80–96.

Clemente, F. M., Couceiro, M. S., Martins, F. M. L., et al. (2015c). Soccer team's tactical behaviour: Measuring territorial domain. *Proceedings of the Institution of Mechanical Engineers, Part P: Journal of Sports Engineering and Technology, 229*(1), 58–66.

Clemente, F. M., Couceiro, M. S., et al. (2015d). Using Network Metrics in Soccer: A Macro-Analysis. *Journal of Human Kinetics, 45*, 123–134.

Costa, I. T., et al. (2009). Princípios Táticos do Jogo de Futebol: conceitos e aplicação. *Motriz, 15* (3), 657–668.

Costa, I. T., et al. (2010). Influence of Relative Age Effects and Quality of Tactical Behaviour in the Performance of Youth Football Players. *International Journal of Performance Analysis in Sport, 10*(2), 82–97.

Cotta, C., et al. (2013). A network analysis of the 2010 FIFA world cup champion team play. *Journal of Systems Science and Complexity, 26*(1), 21–42.

Couceiro, M. S., Clemente, F. M., & Martins, F. M. L. (2013). Analysis of football player's motion in view of fractional calculus. *Central European Journal of Physics, 11*(6), 714–723.

Coutts, A. J. (2014). Evolution of football match analysis research. *Journal of sports sciences, 32*(20), 1829–1830. Available at: http://www.ncbi.nlm.nih.gov/pubmed/25494396. Accessed December 16, 2014.

Duarte, R., et al. (2012). Sports teams as superorganisms: Implications of sociobiological models of behaviour for research and practice in team sports performance analysis. *Sports Medicine, 42*(8), 633–642.

Duch, J., Waitzman, J. S., & Amaral, L. A. (2010). Quantifying the performance of individual players in a team activity. *PLoS ONE, 5*(6), e10937.

Dufour, W. (1993). Computer-assisted scouting in football. *Science and football II* (pp. 160–166). E & FN Spon: London, UK.

Franks, I. M., & McGarry, T. (1996). The science of match analysis. In T. Reilly (Ed.), *Science and football* (pp. 363–375). Oxon: Spon Press Taylor & Francis Group.

Frencken, W., et al. (2011). Oscillations of centroid position and surface area of football teams in small-sided games. *European Journal of Sport Science, 11*(4), 215–223.

Frencken, W., & Lemmink, K. (2008). Team kinematics of small-sided football games: A systematic approach. In T. Reilly & F. Korkusuz (Eds.), *Science and football VI* (pp. 161–166). Routledge, Taylor & Francis Group: Oxon.

Frencken, W. G. P., Lemmink, K. A. P. M., & Delleman, N. J. (2010). Soccer-specific accuracy and validity of the local position measurement (LPM) system. *Journal of science and medicine in sport/ Sports Medicine Australia, 13*(6), 641–645.

Grant, A. G., Williams, A. M., & Reilly, T. (1999). Analysis of goals scored in the 1998 World Cup. *Journal of Sports Sciences, 17*, 826–827.

Gréhaigne, J. F. (1992). *L'organisation du jeu en football.* Joinville-le-Pont, France: Éditions Actio.

Gréhaigne, J. F., Bouthier, D., & David, B. (1997). Dynamic-system analysis of opponent relationship in collective actions in football. *Journal of Sports Sciences, 15*(2), 137–149.

Gréhaigne, J. F., Godbout, P., & Bouthier, D. (1999). The foundations of tactics and strategy in team sports. *Journal of Teaching in Physical Education, 18*, 159–174.

Gréhaigne, J. F., Godbout, P., & Zerai, Z. (2011). How the "rapport de forces" evolves in a football match: the dynamics of collective decisions in a complex system. *Revista de Psicología del Deporte, 20*(2), 747–765.

Gréhaigne, J. F., Marchal, D., & Duprat, E. (2001). Regaining possession of the ball in the defense area in football. In W. Spinks, T. Reilly, & A. Murphy (Eds.), *Revue sciences and football.* London, UK: Taylor & Francis.

Grund, T. U. (2012). Network structure and team performance: The case of English Premier League soccer teams. *Social Networks, 34*(4), 682–690.

Horvath, S. (2011). *Weighted network analysis: Applications in genomics and systems biology.* New York: Springer.

Hughes, M. D., & Bartlett, R. M. (2002). The use of performance indicators in performance analysis. *Journal of Sports Sciences, 20*(10), 739–754.

Hughes, M., & Churchill, S. (2005). Attacking profiles of successful and unsuccessful teams in Copa America 2001. In T. Reilly, J. Cabri & D. Araújo (Eds.) *Science and football V* (pp. 222–228). London, UK: Routledge, Taylor & Francis Group.

Hughes, M., & Franks, M. (2004). *Notational analysis of sport.* London, UK: Routledge.

Hughes, M., & Franks, I. (2005). Analysis of passing sequences, shots and goals in soccer. *Journal of Sports Sciences, 23*(5), 509–514.

Jinshan, X., et al. (1993). Analysis of the goals in the 14th World Cup. In T. Reilly, J. Clarys, & A. Stibbe (Eds.), *Science and football II* (pp. 203–205). E. & F. Spon: London, UK.

Jones, P. D., James, N., & Mellallieu, S. D. (2004). Possession as a performance indicator in footbal. *International Journal of Performance Analysis of Sport, 4*, 98–102.

Jonsson, G. K., et al. (2006). Hidden patterns of play interaction in soccer using SOF-CODER. *Behavior Research Methods, 38*(3), 372–381.

Lago, C. (2009). The influence of match location, quality of opposition, and match status on possession strategies in professional association football. *Journal of Sports Sciences, 27*(13), 1463–1469.

Lago, C., & Martín, R. (2007). Determinants of possession of the ball in football. *Journal of Sports Sciences, 25*(9), 969–974.

Lago-Peñas, C., & Dellal, A. (2010). Ball possession strategies in elite soccer according to the evolution of the match-score: The influence of situational variables. *Journal of Human Kinetics, 25*, 93–100.

Landis, J. R., & Koch, G. C. (1977). The measurement of observer agreement for categorical data. *Biometrics, 33*, 159–174.

Lemoine, A., Jullien, H., & Ahmaidi, S. (2005). Technical and tactical analysis of one-touch playing in soccer-Study of the production of information. *International Journal of Performance Analysis in Sport, 5*(1), 83–103.

Lusher, D., Robins, G., & Kremer, P. (2010). The application of social network analysis to team sports. *Measurement in Physical Education and Exercise Science, 14*(4), 211–224.

Magnusson, M. S. (2000). Discovering hidden time patterns in behavior: T-patterns and their detection. *Behavior Research Methods, Instruments, & Computers, 32*(1), 93–110.

Magnusson, M. S. (1996). Hidden real-time patterns in intra-and inter-individual behavior: Description and detection. *European Journal of Psychological Assessment, 12*(2), 112–123.

Malta, P., & Travassos, B. (2014). Characterization of the defense-attack transition of a soccer team. *Motricidade, 10*(1), 27–37.

McGarry, T. (2005). Soccer as a dynamical system: some theoretical considerations. In T. Reilly, J. Cabri & D. Araújo, (Eds.), *Science and football V* (pp. 570–579). London and New York: Routledge, Taylor & Francis Group.

Moura, F. A., et al. (2012). Quantitative analysis of Brazilian football players' organization on the pitch. *Sports Biomechanics, 11*(1), 85–96.

Okihara, K., et al. (2004). Compactness as a strategy in a football match in relation to a change in offense and defence. *Journal of Sports Sciences, 22*(6), 515.

Passos, P., et al. (2011). Networks as a novel tool for studying team ball sports as complex social systems. *Journal of Science and Medicine in Sport, 14*(2), 170–176.

Peña, J. L., & Touchette, H. (2012). A network theory analysis of football strategies. In *arXiv preprint arXiv* (p. 1206.6904).

Rampinini, E. et al. (2014). Accuracy of GPS devices for measuring high-intensity running in field-based team sports. *International Journal of Sports Medicine*.

Reep, C., & Benjamin, B. (1968). Skill and chance in Association Football. *Journal of the Royal Statistical Society, Series A (General), 131*(4), 581–585.

Reep, C., Pollard, R., & Benjamin, B. (1971). Skill and chance in ball games. *Journal of the Royal Statistical Society Series A (General), 134*, 623–629.

Robinson, G., & O'Donoghue, P. (2007). A weighted kappa statistic for reliability testing in performance analysis of sport. *International Journal of Performance Analysis in Sport, 7*(1), 12–19.

Sarmento, H., et al. (2014). Match analysis in football: A systematic review. *Journal of Sports Sciences, 32*(20), 1831–1843.

Sarmento, H., et al. (2014b). Match analysis in football: a systematic review. *Journal of sports sciences*, 1–13.

Seabra, F. (2010). *Identificação e análise de padrões de circulação da bola no futebol.* Unpublished Thesis. São Paulo, Brasil: Universidade de São Paulo.

Tabachnick, B., & Fidell, L. (2007). *Using multivariate statistics.* New York, USA: Harper & Row Publishers.

Taylor, J. B., et al. (2008). The influence of match location, quality of opposition, and match status on technical performance in professional association football. *Journal of Sports Sciences, 26*(9), 885–895.

Tenga, A., et al. (2010). Effect of playing tactics on achieving score-box possessions in a random series of team possessions from Norwegian professional soccer matches. *Journal of Sports Sciences, 28*(3), 245–255.

Travassos, B., et al. (2013). Performance analysis in team sports: Advances from an ecological dynamics approach. *International Journal of Performance Analysis in Sport, 13*(1), 83–95.

Vilar, L., et al. (2013). Science of winning football: emergent pattern-forming dynamics in association football. *Journal of Systems Science and Complexity, 26*, 73–84.

Chapter 4
Micro Levels of Analysis: Player's Centralities in the Team

Abstract The prominence level of a player in the team can be measured in social network analysis. For that reason, the aim of this chapter is to present the centrality metrics that can be applied in team sports analysis. The presentation will also allow identifying the specific formulas for weighted and unweighted graphs and digraphs. Finally, a brief interpretation for team sports analysis will be provided.

Keywords Social network analysis · Graph theory · Digraphs · Centrality measures · Micro-analysis

4.1 Degree Centrality

Definition 4.1 (Wasserman and Faust 1994) Let n_i be a vertex of unweighted graph G with n vertices. The degree centrality index, $C_D = (n_i)$, of a vertex is calculated as

$$C_D(n_i) = k_i = \sum_{j=1}^{n} a_{ij} = \sum_{j=1}^{n} a_{ji}, \qquad (4.1)$$

where a_{ij} are elements of the adjacency matrix of a G.

Definition 4.2 (Wasserman and Faust 1994) Let n_i be a vertex of unweighted graph G with n vertices. The standardized degree centrality index, $C'_{(D)}(n_i)$, is the proportion the vertices that are adjacent to n_i, and is calculated as

$$C'_{(D)}(n_i) = \frac{k_i}{n-1} \qquad (4.2)$$

where k_i is the degree centrality index of the vertex n_i.

© The Author(s) 2016
F.M. Clemente et al., *Social Network Analysis Applied to Team Sports Analysis*,
SpringerBriefs in Applied Sciences and Technology,
DOI 10.1007/978-3-319-25855-3_4

Definition 4.3 (Wasserman and Faust 1994) Let n_i be a vertex of unweighted digraph G with n vertices. The degree centrality index, $C_{D-out}(n_i)$, of the vertex n_i, is calculated as

$$C_{D-out}(n_i) = k_i^{out} = \sum_{j=1}^{n} a_{ij},$$ (4.3)

where a_{ij} are elements of the adjacency matrix of a G.

Definition 4.4 (Wasserman and Faust 1994) Let n_i be a vertex of unweighted digraph G with n vertices. The standardized degree centrality index, $C'_{(D-out)}(n_i)$, is the proportion the vertices that are adjacent to n_i, and is calculated as

$$C'_{(D-out)}(n_i) = \frac{k_i^{out}}{(n-1)^2},$$ (4.4)

where k_i^{out} is the degree centrality index of the vertex n_i.

Definition 4.5 (Opsahl et al. 2010) Let n_i be a vertex of weighted graph G with n vertices. The degree centrality index, $C_D^w(n_i)$, of the vertex n_i, is calculated as

$$C_D^w(n_i) = k_i^w = \sum_{j=1}^{n} a_{ij} = \sum_{j=1}^{n} a_{ji},$$ (4.5)

where a_{ij} are elements of the weighted adjacency matrix of a G.

Definition 4.6 (Opsahl et al. 2010) Let n_i be a vertex of weighted graph G with n vertices. The standardized degree centrality index, $C_D^{\prime w}(n_i)$, is the proportion of the weighted of vertices that are adjacent to n_i, and is calculated as

$$C_D^{\prime w}(n_i) = \frac{k_i^w}{\sum_{i=1}^{n} \sum_{\substack{j=1 \\ j \neq 1}}^{n} a_{ij}},$$ (4.6)

where k_i^w is the degree centrality index of the vertex n_i and a_{ij} are elements of the weighted adjacency matrix of a G.

Definition 4.7 (Opsahl et al. 2010) Let n_i be a vertex of weighted digraph G with n vertices. The degree centrality index, $C_D^w(n_i)$, of the vertex n_i, is calculated as

$$C_{D-out}^w(n_i) = k_i^{w-out} = \sum_{j=1}^{n} a_{ij},$$ (4.7)

where a_{ij} are elements of the weighted adjacency matrix of a G.

Definition 4.8 (Opsahl et al. 2010) Let n_i be a vertex of weighted digraph G with n vertices. The standardized degree centrality index, $C'^w_{(D-out)}(n_i)$, is the proportion of the weight of vertices that are adjacent to n_i, and is calculated as

$$C'^w_{(D-out)}(n_i) = \frac{k_i^{w-out}}{\sum_{i=1}^{n}\sum_{\substack{j=1\\j\neq i}}^{n} a_{ij}}, \tag{4.8}$$

where k_i^{w-out} is the degree centrality index of the vertex n_i and a_{ij} are elements of the weighted adjacency matrix of a G.

4.1.1 Team Sports Network Interpretation

Can be interpreted as a measure of the activity of each player (Clemente et al. 2015). Players with higher centrality are connected to more teammates than those with lower centrality. Thus, such nodes are believed to be more important for the overall network structure.

In the context of the soccer and analysis to passes signifies that the players with larger centrality scores are those who contributed more to their team's offensive attempts through their passes to the other players of their team.

4.2 Closeness Centrality

Definition 4.9 (Opsahl et al. 2010; Rubinov and Sporns 2010) Given two vertices n_i and n_j of the unweighted graph G with n vertices. The geodesic distance between n_i and n_j is obtained by

$$d(n_i, n_j) = \min_{ijh}(a_{ih} + \cdots + a_{hj}) \tag{4.9}$$

where h are intermediary vertices on paths between vertices n_i and n_j, and a_{uv} are the elements of the adjacency matrix of the G.

Remark 4.1 The geodesic distance between n_i and n_j in unweighted digraphs is determined in the similar form that unweighted graphs.

Definition 4.10 (Wasserman and Faust 1994) Given an unweighted graph G with n vertices. The closeness index of the n_i, $C_{(C)}(n_i)$, is determined by

$$C_{(C)}(n_i) = \left[\sum_{\substack{j=1 \\ i \neq j}}^{n} d(n_i, n_j) \right]^{-1}, \tag{4.10}$$

where $d(n_i, n_j)$ is a geodesic distance between n_i and n_j.

Remark 4.2 The closeness of one vertex n_i of unweighted digraph is determined in the similar form that unweighted graph.

Definition 4.11 (Wasserman and Faust 1994) Given an unweighted graph G with n vertices. The standardized closeness index of the vertex n_i is obtained by

$$C'_{(C)}(n_i) = (n - 1) \times \left[\sum_{\substack{j=1 \\ i \neq j}}^{n} d(n_i, n_j) \right]^{-1}, \tag{4.11}$$

where $d(n_i, n_j)$ is a geodesic distance between n_i and n_j.

Remark 4.3 The standardized closeness of one vertex n_i of unweighted digraph is determined in the similar form that unweighted graph.

Definition 4.12 (Opsahl et al. 2010) Given two vertices n_i and n_j of the weighted graph G with n vertices. The geodesic distance between n_i and n_j is obtained by

$$d^w(n_i, n_j) = \min_{ijh} \left(\frac{1}{a_{ih}} + \cdots + \frac{1}{a_{hj}} \right) \tag{4.12}$$

where h are intermediary vertices on paths between vertices n_i and n_j, and a_{uv} are the elements of the weighted adjacency matrix of the G.

Remark 4.4 The geodesic distance between n_i and n_j in weighted digraphs is determined in the similar form that weighted graphs.

Definition 4.13 (Opsahl et al. 2010) Given a weighted graph G with n vertices. The closeness index of the n_i, $C^w_{(C)}(n_i)$, is determined by

$$C^w_{(C)}(n_i) = \left[\sum_{\substack{j=1 \\ i \neq j}}^{n} d^w(n_i, n_j) \right]^{-1}, \tag{4.13}$$

where $d^w(n_i, n_j)$ is a geodesic distance between n_i and n_j.

Remark 4.5 The closeness of one vertex n_i of weighted digraph is determined in the similar form that weighted graph.

Definition 4.14 (Rubinov and Sporns 2010; Opsahl et al. 2010) Given a weighted graph G with n vertices. The standardized closeness index of the vertex n_i is obtained by

$$C_{(C)}^{\prime w}(n_i) = (n-1) \times \left[\sum_{\substack{j=1 \\ i \neq j}}^{n} d^w(n_i, n_j) \right]^{-1}, \tag{4.14}$$

where $d^w(n_i, n_j)$ is a geodesic distance between n_i and n_j.

Remark 4.6 The standardized closeness of one vertex n_i of weighted digraph is determined in the similar form that weighted graph.

4.2.1 Team Sports Network Interpretation

In any case, the centrality score of a player quantifies the proximity of how close is such player to its peers. Players with higher centrality scores can reach more players in fewer passes than those with lower centrality scores. This measure can also be interpreted as an index of the capability of a player to access or pass information to other players in the network (Clemente et al. 2015).

In the context of passes between teammates, the centrality index of each player denotes how close, in terms of passes, that player has been to all other teammates during the development of the team's attack. High centrality scores of a player might indicate that this player not only participated in the attacks successfully passing to other players, but was also closer to the final outcome of the attack (i.e., shoot and out) (Clemente et al. 2015).

4.3 Stress Centrality

Definition 4.15 (Brandes 2001) Given an unweighted graph $G = (V, E)$, with $n_i, n_j, n_k \in V$, $i, j, k = 1, \ldots, n$. The stress centrality index is calculated by:

$$C_b(n_k) = \sum_{\substack{n_i, n_j \in V \\ i \neq nj \neq k}} g_{ij}(n_k), \qquad (4.15)$$

where $g_{ij}(n_k)$ is the number of shortest paths between n_i and n_j that pass through n_k.

Remark 4.7 The stress centrality index in weighted or/and direct graphs, is determined the similar form that unweighted graphs shut that the path lengths are obtained on respective weighted or direct paths.

4.3.1   Team Sports Network Interpretation

The stress of a node in team sports analysis, for instance the passes to teammates, can indicate the relevance of a player as functionally capable of holding together interactive players. The higher the value the higher the relevance of the player in connecting attacking plays. Due to the nature of this centrality, it is possible that the stress simply indicates a player heavily involved in attacking processes but not relevant to maintain the interaction between other player.

4.4   Betweenness Centrality

Definition 4.16 (Opsahl et al. 2010) Given a unweighted graph $G = (V, E)$, with $n_i, n_j, n_k \in V$, $i, j, k = 1, \ldots, n$. The betweenness centrality index is calculated by:

$$C_b(n_k) = \sum_{\substack{n_i, n_j \in V \\ i \neq nj \neq k}} \frac{g_{ij}(n_k)}{g_{ij}}, \qquad (4.16)$$

where $g_{ij}(n_k)$ is the number of shortest paths between n_i and n_j that pass through n_k and g_{ij} is the number of shortest paths between n_i and n_j.

Definition 4.17 (Rubinov and Sporns 2010) Given an unweighted graph $G = (V, E)$, with $n_i, n_j, n_k \in V$, $i, j, k = 1, \ldots, n$. The standardized betweenness centrality index is calculated by:

$$C_b'(n_k) = \frac{1}{(n-1)(n-2)} \sum_{\substack{n_i, n_j \in V \\ i \neq nj \neq k}} \frac{g_{ij}(n_k)}{g_{ij}}, \qquad (4.17)$$

where $g_{ij}(n_k)$ is the number of shortest paths between n_i and n_j that pass through n_k and g_{ij} is the number of shortest paths between n_i and n_j.

Remark 4.8 The betweenness centrality and standardized betweenness centrality in weighted or/and direct graphs, is determined the similar form that unweighted graphs shut that the path lengths are obtained on respective weighted or direct paths (Rubinov and Sporns 2010).

4.4.1 Team Sports Network Interpretation

The BC index is often considered the most meaningful measure among other centrality indices because it successfully quantifies how often each player lies between other nodes of the network, perhaps acting as a mediator or "bridge" for them. In essence, the BC score of each player can be explained as a measure of the relative control that player has on other players. In social network analyses, players with higher BC scores are commonly assumed to have a higher probability to exert control on the information flow between other nodes in the same network.

As regards the passing game data, players with higher BC scores might be those who more often were situated between their teammates. For instance, a player with high BC score could be important in passing the ball to others.

4.5 Eccentricity Centrality

Definition 4.18 (Pavlopoulos et al. 2011) Given an unweighted graph G with n vertices The eccentricity centrality, $C_{ecc}(n_i)$, of a vertex n_i is determined by

$$C_{ecc}(n_i) = \frac{1}{\max\{d(n_i, n_j)\}} \tag{4.18}$$

where $d(n_i, n_j)$ is the shortest path between vertices n_i and n_j.

Remark 4.9 The *eccentricity centrality* in weighted or/and direct graphs, is determined the similar form that unweighted graphs shut that the path lengths are obtained on respective weighted or direct paths.

4.5.1　Team Sports Network Interpretation

It can be thought of as how far a player is from the player most distant from it in the network. The eccentricity of a player in a team sports network, for instance the passes to teammates, can be interpreted as the easiness of a player to be reached by all other teammates in the network. Thus, a player with high eccentricity, compared to the average eccentricity of the network, will be more easily influenced by the activity of other teammates or, conversely could easily influence several other teammates. In contrast, a low eccentricity, compared to the average eccentricity of the network, could indicate a marginal functional role.

4.6　Degree Prestige

Definition 4.19 (Wasserman and Faust 1994) Let n_i be a vertex of unweighted graph G with n vertices. The degree prestige index, $P_D(n_i)$, of a vertex is calculated as

$$P_D(n_i) = k_i = \sum_{j=1}^{n} a_{ij} = \sum_{j=1}^{n} a_{ji}, \tag{4.19}$$

where a_{ij} are elements of the adjacency matrix of a G.

Remark 4.10 When G is a unweighted graph the degree prestige is equal a degree centrality.

Definition 4.20 (Wasserman and Faust 1994) Let n_i be a vertex of unweighted graph G with n vertices. The standardized degree prestige index, $P'_{(D)}(n_i)$, is the proportion the vertices that are adjacent to n_i, and is calculated as

$$P'_{(D)}(n_i) = \frac{k_i}{n-1} \tag{4.20}$$

where k_i is the degree prestige index of the vertex n_i.

Definition 4.21 (Wasserman and Faust 1994) Let n_i be a vertex of unweighted digraph G with n vertices. The degree prestige index, $P_{D-in}(n_i)$, of the vertex n_i, is calculated as

$$P_{D-in}(n_i) = k_i^{in} = \sum_{j=1}^{n} a_{ji}, \tag{4.21}$$

where a_{ij} are elements of the adjacency matrix of a G.

Definition 4.22 (Wasserman and Faust 1994) Let n_i be a vertex of unweighted digraph G with n vertices. The standardized degree prestige index, $P'_{(D-\text{in})}(n_i)$, is the proportion the vertices that are adjacent to n_i is calculated as

$$P'_{(D-\text{in})}(n_i) = \frac{k_i^{in}}{(n-1)^2},$$

(4.22)

where k_i^{in} is the degree prestige index of the vertex n_i.

Definition 4.23 (Opsahl et al. 2010) Let n_i be a vertex of weighted graph G with n vertices. The degree prestige index, $P_D^w(n_i)$, of the vertex n_i, is calculated as

$$P_D^w(n_i) = k_i^w = \sum_{j=1}^{n} a_{ij} = \sum_{j=1}^{n} a_{ji},$$

(4.23)

where a_{ij} are elements of the weighted adjacency matrix of a G.

Remark 4.11 When G is a weighted graph the degree prestige is equal a degree centrality.

Definition 4.24 (Opsahl et al. 2010) Let n_i be a vertex of weighted graph G with n vertices. The standardized degree prestige index, $P_D^{\prime w}(n_i)$, is the proportion of the weighted of vertices that are adjacent to n_i, and is calculated as

$$P_D^{\prime w}(n_i) = \frac{k_i^w}{\sum_{i=1}^{n} \sum_{\substack{j=1 \\ j \neq i}}^{n} a_{ij}},$$

(4.24)

where k_i^w is the degree prestige index of the vertex n_i and a_{ij} are elements of the weighted adjacency matrix of a G.

Definition 4.25 (Opsahl et al. 2010) Let n_i be a vertex of weighted digraph G with n vertices. The degree prestige index, $P_{D-\text{in}}^w(n_i)$, of the vertex n_i, is calculated as

$$P_{D-\text{in}}^w(n_i) = k_i^{w-in} = \sum_{j=1}^{n} a_{ji},$$

(4.25)

where a_{ij} are elements of the weighted adjacency matrix of a G.

Definition 4.26 (Opsahl et al. 2010) Let n_i be a vertex of weighted digraph G with n vertices. The standardized degree prestige index, $P_{(D-\text{in})}^{\prime w}(n_i)$, is the proportion of the weight of vertices that are adjacent to n_i, and is calculated as

$$P'^w_{(D-in)}(n_i) = \frac{k_i^{w-in}}{\sum_{i=1}^n \sum_{\substack{j=1 \\ j \neq i}}^n a_{ij}}, \tag{4.26}$$

where k_i^{w-in} is the degree prestige index of the vertex n_i and a_{ij} are elements of the weighted adjacency matrix of a G.

4.6.1 Team Sports Network Interpretation

It is often used as indication of the "prestige" of each node among its teammates. Players with high centrality scores are those that receive many inbound links from other players. These links can be interpreted as "choices" or "nominations" to a specific actor from others. Thus, a larger centrality score of a player indicates that this player is more prestigious or important among its teammates.

Analysis of the data in passes sequences shows that the players with higher centrality scores are obviously those to whom their teammates preferred to pass the ball more often. These players might possibly be the ones crucial for their team's offensive development because they receive the ball more often than other players during their team's attempt to attack.

4.7 Proximity Prestige

Definition 4.27 (Schramm 2012) Let n_i be a vertex of unweighted digraph G with n vertices. The proximity prestige index, $P_P(n_i)$, of the vertex n_i, is the proportion of vertices who can reach n_i to the average distance these vertices are from n_i, and is determined by

$$P_P(n_i) = \frac{\frac{I_i}{n-1}}{\frac{\sum_{\substack{j=1 \\ j \neq i}}^n d(n_i,n_j)}{I_i}}, \tag{4.27}$$

where I_i is the number the vertices that are either directly or indirectly connected to n_i and $d(n_i, n_j)$ is the shortest path between vertices n_i and n_j.

Remark 4.12 In weighted digraph the proximity prestige index is determined the similar form that unweighted digraphs, considering $d^w(n_i, n_j)$.

4.7.1 Team Sports Network Interpretation

In proximity prestige what matters is how close are all the other teammates to a specific player. In the case, this measure will allow identify how important a player is in the team. If a player has a greater proximity of their teammates, this may suggest that the teammates tend to play to him in the case of analysis to passes.

4.8 Eigenvector Centrality

Definition 4.28 (Qi et al. 2012) Let G be a unweighted graph with n vertices. The eigenvector centrality index of the vertex n_i is defined as the ith component of the eigenvector, $\vec{x}$, corresponding to the greatest eigenvalue, of the following characteristic equation $A\vec{x} = \vec{x}\lambda$, where A is the adjacency matrix of G.

Remark 4.12 The eigenvector centrality index of a unweighted digraph is obtained the similar form that unweighted graph.

4.8.1 Team Sports Network Interpretation

Eigenvector allows an immediate evaluation of the regulatory relevance of the player. A player with a very high Eigenvector is a player interacting with several important teammates, thus suggesting a central regulatory role. A player with low Eigenvector, can be considered a peripheral teammate, interacting with few and not central players.

4.9 Subgraph Centrality

Definition 4.29 (Qi et al. 2012) Let G_S be a subgraph of unweighted graph G with n vertices. The subgraph centrality of the vertex n_i is defined as

$$C_{G_S}(n_i) = \sum_{k=0}^{\infty} \frac{u_k(i,i)}{k!}, \tag{4.28}$$

where $u_k(i, i)$ is the number of closed walks with length k that vertex n_i participates in G.

Remark 4.13 The subgraph centrality index of a unweighted digraph is obtained the similar form that unweighted graph.

4.9.1 Team Sports Network Interpretation

In a team sports analysis, subgraph centrality rates the importance of a player based on the number of closed walks beginning and ending at a particular player. These closed walks are weighted based on length, such that the shortest walks contribute the greatest towards the centrality value. The subgraph characterizes the participation of each vertex in all subgraphs in a network, with more weight given smaller subgraphs than larger ones.

4.10 Laplacian Centrality

Definition 4.30 (Qi et al. 2012) Let G be a weighted graph with n vertices. The matrix $L = S - W$ is called the Laplacian matrix of G, where $W = [w_{ij}]$ with $w_{ij} = w_{ji}$ that is the weight of the $(n_i, n_j) \in E$ and $w_{ii} = 0$, and $S = [s_{ij}]$ with $s_{ii} = \sum_{j=1}^{n} w_{ij}$ and $s_{ij} = 0$.

Definition 4.31 (Qi et al. 2012) Let G be a weighted graph with n vertices. The Laplacian energy of G is defined as the following invariant

$$E_L(G) = \sum_{i=1}^{n} \lambda_i^2, \tag{4.29}$$

where λ_i are eigenvalues of the Laplacian matrix of G.

Definition 4.32 (Qi et al. 2012) Given one G weighted graph with n vertices and let G_i be the weighted graph by deleting n_i from G. The Laplacian centrality index, $C_L(n_i)$, of vertex n_i is obtained by

$$C_L(n_i) = \frac{E_L(G) - E_L(G_i)}{E_L(G)}, \tag{4.30}$$

where $E_L(G)$ is the Laplacian energy of G and $E_L(G_i)$ is the Laplacian energy of G_i.

Remark 4.14 The Laplacian centrality index of a weighted graph is not applied the similar form that weighted digraph (Qi et al. 2012).

4.10.1 Team Sports Network Interpretation

Laplacian centrality is an intermediate measuring between global and local characterization of the importance (centrality) of a vertex. The Laplacian centrality of some vertex is actually related to the number of 2-walks it participates in. That is, it not only takes into account the local environment immediately around it but also a bigger environment around its neighbours. It is an intermediate between global and local characterization of the position of a player in weighted networks.

4.11 Directed Information Centrality

Definition 4.33 (Poulakakis et al. 2015) Given one G weighted digraph with n vertices. The matrix Laplacian matrix, L_D, associated with G is defined by

$$(L_D)_{ij} := \begin{cases} \sum_{i=1, i \neq j}^{n} w_{ij}, & i = j \\ -w_{ij}, & i \neq j \end{cases} \tag{4.31}$$

where w_{ij} is element of weighted adjacency matrix of G.

Definition 4.34 (Poulakakis et al. 2015) Given one G weighted digraph with n vertices. Let $n_i, n_j \in V$ be two nodes with $n_i \neq n_j$. Then, the directed effective resistance between n_i and n_j is calculated by

$$r(n_i, n_j) = x_{ii} + x_{jj} - 2x_{ij}, \tag{4.32}$$

where x_{ij}, $1 \leq i, j \leq n$ are the elements of the matrix $X := 2Q^T \Sigma Q$, and Q is the $(n-1) \times n$ matrix with rows that an orthonormal basis for this subspace and satisfies

$$Q1_n = O_{n-1}, \ QQ^T = I_{n-1} \quad and \quad QQ^T = I_n - \frac{1}{n} 1_n 1_n^T, \tag{4.33}$$

and Σ is solution of equation

$$L_{D_r} \Sigma + \Sigma L_{D_r} = I_{n-1}, \tag{4.34}$$

where $L_{D_r} = QL_DQ^T$, 1_n is n-dimensional vector with all entries equal to one and I_n is the $n \times n$ identity matrix.

Definition 4.35 (Poulakakis et al. 2015) Given one G weighted digraph with n vertices. The directed information centrality index of a vertex n_i as the inverse of mean effective resistance $r(n_i, n_j)$ over all n_j,

$$DI_C(n_i) = \left(\frac{1}{n} \sum_{j=1}^{n} r(n_i, n_j) \right)^{-1}. \tag{4.35}$$

4.11.1 Team Sports Network Interpretation

It is based on information that can be transmitted between any two points in a connected network (Estrada and Hatano 2010). The path connecting two players is considered as a "signal", while the "noise" in the transmission of the signal is measured by the variance of this signal (Estrada and Hatano 2010). In the case of passes, it can be considered to analyze the capacity to move the ball for their teammates.

4.12 PageRank Centrality

Definition 4.36 (Poulakakis et al. 2015) Given one G unweighted digraph with n vertices. The pagerank centrality index, $PR(n_i)$, of a vertex n_i is determined by

$$PR(n_i) = \mathrm{p} \sum_{j \neq i} \frac{a_{ji}}{k_j^{out}} PR(n_j) + q, \tag{4.36}$$

where a_{ji} are elements of the adjacency matrix p is a heuristic parameter representing the probability of this vertex connect with others and q is a parameter awarding a "free" popularity to each vertex.

Remark 4.15 The pagerank centrality index of a weighted digraph is applied the similar form that unweighted digraph.

4.12.1 Team Sports Network Interpretation

Pagerank centrality is a recursive notion of 'popularity' or importance which follows the principle that 'a player is popular if he gets passes from other popular players'. Pagerank centrality roughly assigns to each player the probability that he will have the ball after a reasonable number of passes has been made. In this metric, probability p can be replaced by player-dependent probabilities pi, which would make more sense if certain players are more tendencies to keep the ball than others. In either case, the value of p (or the pi's) does not come from the network alone, as it might in general be very different from one team to another, and should be determined by heuristics.

4.13   Power Centrality

Definition 4.37 (Borgatti and Everett 2006) Given one G unweighted graph with n vertices. The power centrality index, $P_C(n_i)$, of a vertex n_i is determined by

$$w_{ij} = \delta \sum_{m=1}^{\infty} b^m \left(a^{m+1} \right)_{ij} \tag{4.37}$$

$$P_C(n_i) = \sum_j w_{ij} \tag{4.38}$$

where a_{ij} are elements of the adjacency matrix of G and δ, b are constants.

Remark 4.16 The power centrality index is applied the similar unweighted digraph, weighted graphs and weighted digraphs.

4.13.1   Team Sports Network Interpretation

The degree allows an immediate evaluation of the regulatory relevance of the node. It is a function of the connections of the players in one's teammates. The more connections the players in your neighbourhood have, the more central they are. The fewer the connections the players in your neighbourhood, the more powerful they are. Therefore, a player that interact with more teammates are well positioned to be considered as more relevant during the network activity.

4.14   Centroid

Definition 4.38 (Scardoni and Laudanna 2012) Given one G unweighted graph with n vertices. The centroid centrality index $C_{Ce}(n_i)$, of a vertex n_i is determined by

$$C_{Ce}(n_i) = \min\{f(n_i, n_j) : n_j \in V - \{n_i\}\} \tag{4.39}$$

where $f(n_i, n_j) = \gamma_{n_i}(n_j) - \gamma_{n_j}(n_i)$, and $\gamma_{n_i}(n_j)$ is the number of vertex closer to n_i than to n_j, i.e. $\gamma_{n_i}(n_j) = |\{n_k \in V : d(n_i, n_k) < d(n_j, n_k)\}|$.

Remark 4.17 (Scardoni and Laudanna 2012) The centroid centrality index is applied the similar unweighted digraph, weighted graphs and weighted digraphs.

4.14.1 Team Sports Network Interpretation

The centrality value of a node in team sports analysis, for instance the passes to teammates, can be interpreted as the probability of a player to be functionally capable of organizing clusters in the team. Thus, a player with high centroid value, compared to the average centroid value of the network, will be possibly involved in coordinating the activity of other highly connected players, altogether devoted to the regulation of team play. Accordingly, a network with a very high average centroid value is more likely organizing functional units, whereas a pass network with very low average centroid value will behave more likely as an open cluster of players connecting different regulatory clusters.

4.15 Clustering Coefficient

Definition 4.39 (Fagiolo 2007; Rubinov and Sporns 2010) Let G be unweighted graph with n vertices. The clustering coefficient index, $CL(n_i)$, of a vertex n_i

$$CL(n_i) = \frac{\sum_{h \neq i, j} a_{ij} a_{ih} a_{jh}}{k_i(k_i - 1)} = \frac{(A^3)_{ii}}{k_i(k_i - 1)} \tag{4.40}$$

where a_{ij} are elements of the adjacency matrix of a G and k_i is degree centrality index.

Definition 4.40 (Fagiolo 2007; Rubinov and Sporns 2010) Let G be unweighted digraph with n vertices. The clustering coefficient index, $CL_D(n_i)$, of a vertex n_i

$$
\begin{aligned}
CL_D(n_i) &= \frac{\sum_{j,h} (a_{ij} + a_{ji})(a_{ih} + a_{hi})(a_{jh} + a_{hj})}{2\left[(k_i^{out} + k_i^{in})(k_i^{out} + k_i^{in} - 1) - 2\sum_{j \neq i} a_{ij} a_{ji}\right]} \\
&= \frac{(A + A^T)_{ij}^3}{2\left[(k_i^{out} + k_i^{in})(k_i^{out} + k_i^{in} - 1) - 2\sum_{j \neq i} a_{ij} a_{ji}\right]}
\end{aligned}
\tag{4.41}
$$

where a_{ij} are elements of the adjacency matrix of a G and k_i^{out} is outdegree index of vertex n_i and k_i^{in} is indegree index of vertex n_i.

Definition 4.41 (Fagiolo 2007; Rubinov and Sporns 2010) Let G be weighted graph with n vertices. The clustering coefficient index, $CL^w(n_i)$, of a vertex n_i

$$CL^w(n_i) = \frac{\sum_{h \neq i, j} a_{ij}^{\frac{1}{3}} a_{ih}^{\frac{1}{3}} a_{jh}^{\frac{1}{3}}}{k_i(k_i - 1)} = \frac{\left(A^{\left[\frac{1}{3}\right]}\right)_{ii}^3}{k_i(k_i - 1)} \tag{4.42}$$

where a_{ij} are elements of the weighted adjacency matrix of a G and k_i is degree centrality index and we define $A^{[\frac{1}{3}]} = \left[a_{ij}^{\frac{1}{3}}\right]$, thus the matrix obtained from A by taking the 3rd root of each entry.

Definition 4.42 (Fagiolo 2007; Rubinov and Sporns 2010) Let G be weighted digraph with n vertices. The clustering coefficient index, $CL_D^w(n_i)$, of a vertex n_i

$$
CL_D^w(n_i) = \frac{\sum_{j,h}\left(a_{ij}^{\frac{1}{3}}+a_{ji}^{\frac{1}{3}}\right)\left(a_{ih}^{\frac{1}{3}}+a_{hi}^{\frac{1}{3}}\right)\left(a_{jh}^{\frac{1}{3}}+a_{hj}^{\frac{1}{3}}\right)}{2\left[(k_i^{out}+k_i^{in})(k_i^{out}+k_i^{in}-1)-2\sum_{j\neq i}a_{ij}a_{ji}\right]},
$$

$$
= \frac{\left[A^{[\frac{1}{3}]}+(A^T)^{[\frac{1}{3}]}\right]_{ii}^3}{2\left[(k_i^{out}+k_i^{in})(k_i^{out}+k_i^{in}-1)-2\sum_{j\neq i}a_{ij}a_{ji}\right]}
$$

(4.43)

where a_{ij} are elements of the weighted adjacency matrix of a G, k_i^{out} is outdegree index of vertex n_i, k_i^{in} is indegree index of vertex n_i and $A^{[\frac{1}{3}]} = \left[a_{ij}^{\frac{1}{3}}\right]$.

4.15.1 Team Sports Network Interpretation

Clustering coefficient quantifies how close a player and its teammates in a graph are to become a clique (a complete subgraph). It can be used to determine whether a graph is a small-world network (a network of small average distance but relative large number of cliques). In teams sports it is possible to observe a higher average clustering coefficient compared to random networks, which proves their clustering nature. In fact, many players' processes are governed by specific interactions that lead to cluster in the team.

References

Borgatti, S. P., & Everett, M. G. (2006). A graph-theoretic perspective on centrality. *Social Networks, 28*(4), 466–484.

Brandes, U. (2001). A faster algorithm for betweenness centrality*. *The Journal of Mathematical Sociology, 25*(2), 163–177.

Clemente, F. M., et al. (2015). Midfielder as the prominent participant in the building attack: A network analysis of national teams in FIFA World Cup 2014. *International Journal of Performance Analysis in Sport, 15*(2), 704–722.

Estrada, E., & Hatano, N. (2010). Resistance distance, information centrality, node vulnerability and vibrations in complex networks. *Network science* (pp. 13–29). London: Springer.

Fagiolo, G. (2007). Clustering in complex directed networks. *Physical Review E, 76*(2), 026107.

Opsahl, T., Agneessens, F., & Skvoretz, J. (2010). Node centrality in weighted networks: Generalizing degree and shortest paths. *Social Networks, 32*(3), 245–251.

Pavlopoulos, G. A., et al. (2011). Using graph theory to analyze biological networks. *BioData mining, 4*(1), 10.

Poulakakis, I. et al. (2015). Information centrality and ordering of nodes for accuracy in noisy decision-making networks. *IEEE Transactions on Automatic Control*, 1–1.

Qi, X., et al. (2012). Laplacian centrality: A new centrality measure for weighted networks. *Information Sciences, 194*, 240–253.

Rubinov, M., & Sporns, O. (2010). Complex network measures of brain connectivity: Uses and interpretations. *NeuroImage, 52*(3), 1059–1069.

Scardoni, G., & Laudanna, C. (2012). *Centralities based analysis of complex networks*. INTECH Open Access Publisher.

Schramm, H. J. (2012). *Freight forwarder's intermediary role in multimodal transport chains: A social network approach*. Vienna, Austria: Springer.

Wasserman, S., & Faust, K. (1994). *Social network analysis: Methods and applications*. New York, USA: Cambridge University Press.

Chapter 5
Meso Level of Analysis: Subgroups in Teams

Abstract In team sports the teammates cooperates with interdependency between them. One of the examples it is the forward player that depend from the backward players to receive the ball. For that reason, it is important to understand how teammates cooperate and autonomous or dependent are a specific player. Moreover, using such idea will be also possible to identify some clusters and patterns of interactions inside the team. For that reason, the aim of this chapter is to present the network measurements that allows to perform a meso-analysis and identify the specific interactions between teammates.

Keywords Social network analysis · Graph theory · Digraphs · Neighbors activity · Meso-analysis

5.1 Average Neighbor Degree

Definition 5.1 (Rubinov and Sporns 2010) Given one unweighted graph G with n vertices. The average neighbors degree index $\overline{ND}(n_i)$ of a vertex n_i, of G is calculated as

$$\overline{ND}(n_i) = \frac{\sum_{j=1}^{n} a_{ij} k_j}{k_i} \qquad (5.1)$$

where a_{ij} are elements of the adjacency matrix of a G, k_i degree centrality index of vertex n_i.

Definition 5.2 (Sierra 2015) Given one weighted graph G with n vertices. The average neighbors degree index $\overline{ND}^w(n_i)$ of a vertex n_i, of G is calculated as

$$\overline{ND}^w(n_i) = \frac{\sum_{j=1}^{n} a_{ij} k_j^w}{k_i^w} \qquad (5.2)$$

© The Author(s) 2016
F.M. Clemente et al., *Social Network Analysis Applied to Team Sports Analysis*,
SpringerBriefs in Applied Sciences and Technology,
DOI 10.1007/978-3-319-25855-3_5

where a_{ij} are elements of the weighted adjacency matrix of a G, k_i^w degree centrality index of vertex n_i.

Definition 5.3 (Rubinov and Sporns 2010) Given one unweighted digraph G with n vertices. The average neighbors degree index $\overline{ND}_D(n_i)$ of a vertex n_i, of G is calculated as

$$\overline{ND}_D(n_i) = \frac{\sum_{j=1}^{n} \left(a_{ij} + a_{ji}\right)\left(k_j^{out} + k_j^{in}\right)}{2 \times \left(k_i^{out} + k_i^{in}\right)} \tag{5.3}$$

where a_{ij} are elements of the adjacency matrix of a G, k_i^{out} and k_i^{in} degree index of vertex n_i.

Definition 5.4 (Rubinov and Sporns 2010) Given one weighted digraph G with n vertices. The average neighbors degree index $\overline{ND}_D^w(n_i)$ of a vertex n_i, of G is calculated as

$$\overline{ND}_D^w(n_i) = \frac{\sum_{j=1}^{n} \left(a_{ij} + a_{ji}\right)\left(k_j^{w-out} + k_j^{w-in}\right)}{2 \times \left(k_i^{w-out} + k_i^{w-in}\right)} \tag{5.4}$$

where a_{ij} are elements of the weighted adjacency matrix of a G, k_i^{w-out} and k_i^{w-in} degree index of vertex n_i.

5.1.1 Team Sports Network Interpretation

The Average neighbor degree measures the correlation levels between pairs of players. This measure allows identifying the patterns of interactions and the strength of interaction between two players. Can be used to identify patterns of cooperation inside the team and to associate these interactions with macro levels of analysis.

5.2 Assortativity Coefficient

Definition 5.5 (Ciglan et al. 2013) Given one unweighted graph G with n vertices. The assortativity coefficient index, r, of G is calculated as

$$r = \cfrac{\cfrac{\sum_{(n_i,n_j)\in E} k_i k_j}{\sum_{i=1}^{n}\sum_{\substack{j=1\\j>i}}^{n} a_{ij}} - \left[\cfrac{\sum_{(n_i,n_j)\in E}\frac{1}{2}(k_i+k_j)}{\sum_{i=1}^{n}\sum_{\substack{j=1\\j>i}}^{n} a_{ij}}\right]^2}{\cfrac{\sum_{(n_i,n_j)\in E}\frac{1}{2}(k_i^2+k_j^2)}{\sum_{i=1}^{n}\sum_{\substack{j=1\\j>i}}^{n} a_{ij}} - \left[\cfrac{\sum_{(n_i,n_j)\in E}\frac{1}{2}(k_i+k_j)}{\sum_{i=1}^{n}\sum_{\substack{j=1\\j>i}}^{n} a_{ij}}\right]^2} \tag{5.5}$$

where a_{ij} are elements of the adjacency matrix of a G, k_i degree index of vertex n_i.

Definition 5.6 (Schott and Wunderlich 2014) Given one weighted graph G with n vertices. The weighted assortativity coefficient index, r^w, of G is calculated as

$$r^w = \cfrac{\cfrac{\sum_{(n_i,n_j)\in E} a_{ij} k_i^w k_j^w}{\sum_{i=1}^{n}\sum_{\substack{j=1\\j>i}}^{n} a_{ij}} - \left[\cfrac{\sum_{(n_i,n_j)\in E}\frac{1}{2}a_{ij}(k_i^w+k_j^w)}{\sum_{i=1}^{n}\sum_{\substack{j=1\\j>i}}^{n} a_{ij}}\right]^2}{\cfrac{\sum_{(n_i,n_j)\in E}\frac{1}{2}a_{ij}((k_i^w)^2+(k_j^w)^2)}{\sum_{i=1}^{n}\sum_{\substack{j=1\\j>i}}^{n} a_{ij}} - \left[\cfrac{\sum_{(n_i,n_j)\in E}\frac{1}{2}a_{ij}(k_i^w+k_j^w)}{\sum_{i=1}^{n}\sum_{\substack{j=1\\j>i}}^{n} a_{ij}}\right]^2} \tag{5.6}$$

where a_{ij} are elements of the weighted adjacency matrix of a G, k_i^w degree index of vertex n_i.

Definition 5.7 (Rubinov and Sporns 2010) Given one unweighted digraph G with n vertices. The directed assortativity coefficient index, r_D, of G is calculated as

$$r_D = \cfrac{\cfrac{\sum_{(n_i,n_j)\in E} a_{ij} k_i^{out} k_j^{in}}{\sum_{i=1}^{n}\sum_{\substack{j=1\\j\neq i}}^{n} a_{ij}} - \left[\cfrac{\sum_{(n_i,n_j)\in E}\frac{1}{2}a_{ij}(k_i^{out}+k_j^{in})}{\sum_{i=1}^{n}\sum_{\substack{j=1\\j\neq i}}^{n} a_{ij}}\right]^2}{\cfrac{\sum_{(n_i,n_j)\in E}\frac{1}{2}a_{ij}((k_i^{out})^2+(k_j^{in})^2)}{\sum_{i=1}^{n}\sum_{\substack{j=1\\j\neq i}}^{n} a_{ij}} - \left[\cfrac{\sum_{(n_i,n_j)\in E}\frac{1}{2}a_{ij}(k_i^{out}+k_j^{in})}{\sum_{i=1}^{n}\sum_{\substack{j=1\\j\neq i}}^{n} a_{ij}}\right]^2} \tag{5.7}$$

where a_{ij} are elements of the adjacency matrix of a G, k_i^{out} and k_i^{in} degree index of vertex n_i.

Definition 5.8 (Schott and Wunderlich 2014) Given one weighted digraph G with n vertices. The weighted directed assortativity coefficient index, r_D^w, of G is calculated as

$$r_D^w == \cfrac{\cfrac{\sum_{(n_i,n_j)\in E} a_{ij} k_i^{w-out} k_j^{w-in}}{\sum_{i=1}^{n}\sum_{\substack{j=1\\j\neq i}}^{n} a_{ij}} - \left[\cfrac{\sum_{(n_i,n_j)\in E}\frac{1}{2} a_{ij}(k_i^{w-out} + k_j^{w-in})}{\sum_{i=1}^{n}\sum_{\substack{j=1\\j\neq i}}^{n} a_{ij}}\right]^2}{\cfrac{\sum_{(n_i,n_j)\in E}\frac{1}{2} a_{ij}((k_i^{w-out})^2 + (k_j^{w-in})^2)}{\sum_{i=1}^{n}\sum_{\substack{j=1\\j\neq i}}^{n} a_{ij}} - \left[\cfrac{\sum_{(n_i,n_j)\in E}\frac{1}{2} a_{ij}(k_i^{w-out} + k_j^{w-in})}{\sum_{i=1}^{n}\sum_{\substack{j=1\\j\neq i}}^{n} a_{ij}}\right]^2} \tag{5.8}$$

where a_{ij} are elements of the adjacency matrix of a G, k_i^{w-out} and k_i^{w-in} degree index of vertex n_i.

Remark 5.1 We also can calculate the (weighted) assortativity coefficient for the indegree and outdegree of (weighted) digraphs (Schott and Wunderlich 2014; Piraveenan et al. 2012).

5.2.1 Team Sports Network Interpretation

Assortativity is often operationalized as a correlation between two players. A network is called assortative if the players with higher degree have the tendency to connect with other players that also have high degree of connectivity (Pavlopoulos et al. 2011). If the players with higher degree have the tendency to connect with other players with low degree then the network is called disassortative (Pavlopoulos et al. 2011).

5.3 Topological Overlap

Definition 5.9 (Horvath 2011) Given one unweighted graph G with n vertices. The topological overlap measure index, TOM_{ij}, between vertices n_i and n_j is calculated as

$$\text{TOM}_{ij} = \begin{cases} \dfrac{\sum_{l\neq i,j} a_{il}a_{jl} + a_{ij}}{\min\left\{\sum_{l\neq i} a_{il} - a_{ij}, \sum_{l\neq j} a_{jl} - a_{ij}\right\} + 1} & i \neq j \\ 1 & i = j \end{cases} \tag{5.9}$$

where a_{ij} are elements of the adjacency matrix of a G.

Remark 5.2 If $0 \leq a_{ij} \leq 1$ then $0 \leq \text{TOM}_{ij} \leq 1$ (Li and Horvath 2007).

Definition 5.10 (Schott and Wunderlich 2014) Given one unweighted digraph G with n vertices. The overlap similarity index between vertices n_i and n_j at incoming or outgoing is calculated as, respectively

$$OS_D^{out}(n_i, n_j) = \frac{|N^{out}(n_i) \bigcap N^{out}(n_j)|}{\min\{|N^{out}(n_i)|, |N^{out}(n_j)|\}} \tag{5.10}$$

$$OS_D^{in}(n_i, n_j) = \frac{|N^{in}(n_i) \bigcap N^{in}(n_j)|}{\min\{|N^{in}(n_i)|, |N^{in}(n_j)|\}} \tag{5.11}$$

where $|N^{out}(n_i)|$ is the number of all outgoing neighbors of n_i and $|N^{in}(n_i)|$ is the number of all incoming neighbors of n_i.

Definition 5.11 (Schott and Wunderlich 2014) Given one weighted graph G with n vertices. The weighted overlap similarity index, OS^w, between vertices n_i and n_j is calculated as, respectively

$$OS^w(n_i, n_j) = \frac{\sum_{n_k \in N(n_i) \bigcap N(n_j)} \min(w_{ik}, w_{jk})}{\min\left\{\sum_{n_k \in N(n_i)} w_{ik}, \sum_{n_k \in N(n_j)} w_{jk}\right\}} \tag{5.12}$$

where w_{ik} is the weight of $(n_i, n_k) \in E$ and $N(n_i)$ is a set of all neighbors of n_i.

Definition 5.12 (Schott and Wunderlich 2014) Given one weighted digraph G with n vertices. The weighted overlap similarity index between vertices n_i and n_j at incoming or outgoing is calculated as, respectively

$$OS_D^{w-out}(n_l, n_j) = \frac{\sum_{n_k \in N^{out}(n_i) \bigcap N^{out}(n_j)} \min(w_{ik}, w_{jk})}{\min\left\{\sum_{n_k \in N^{out}(n_i)} w_{ik}, \sum_{n_k \in N^{out}(n_j)} w_{jk}\right\}} \tag{5.13}$$

$$OS_D^{w-in}(n_i, n_j) = \frac{\sum_{n_k \in N^{in}(n_i) \bigcap N^{in}(n_j)} \min(w_{ki}, w_{kj})}{\min\left\{\sum_{n_k \in N^{in}(n_i)} w_{ki}, \sum_{n_k \in N^{in}(n_j)} w_{kj}\right\}} \tag{5.14}$$

where w_{ki} is the weight of $(n_k, n_i) \in E$.

5.3.1 Team Sports Network Interpretation

The topological overlap measure represents the pair of players that cooperates with the same players (Clemente et al. 2014). Topological overlap measure may also represent the overlap between two players even if they do not participate in the same attacking plays with one another (Clemente et al. 2014). Thus, this measure allows to identify the patterns of interaction between triads without consider the specific pairs relations.

References

Ciglan, M., Laclavík, M., & Nørvåg, K. (2013). On community detection in real-world networks and the importance of degree assortativity. In *Proceedings of the 19th ACM SIGKDD International Conference on Knowledge Discovery and Data Mining* (pp. 1007–1015). ACM.

Clemente, F. M., et al. (2014). Using network metrics to investigate football team players' connections: A pilot study. *Motriz, 20*(3), 262–271.

Horvath, S. (2011). *Weighted network analysis: Applications in genomics and systems biology.* New York: Springer.

Li, A., & Horvath, S. (2007). Network neighborhood analysis with the multi-node topological overlap measure. *Bioinformatics, 23*(2), 222–231.

Pavlopoulos, G. A., et al. (2011). Using graph theory to analyze biological networks. *BioData mining, 4*(1), 10.

Piraveenan, M., Prokopenko, M., & Zomaya, A. (2012). Assortative mixing in directed biological networks. *IEEE/ACM Transactions on Computational Biology and Bioinformatics (TCBB), 9* (1), 66–78.

Rubinov, M., & Sporns, O. (2010). Complex network measures of brain connectivity: Uses and interpretations. *NeuroImage, 52*(3), 1059–1069.

Schott, C., & Wunderlich, M. (2014). *Analyzing live click traces using dynamic graphs.* Technische Universität Darmstadt.

Sierra, D. S. (2015). *Self-organizing cultured neural networks: Image analysis techniques for longitudinal tracking and modeling of the underlying network structure.* Escuela Técnica Superior de Ingenieros de Telecomunicación: Universidad Politécnica de Madrid.

Chapter 6
Macro Levels of Analysis: Network of the Team

Abstract The global properties of a graph may provide important information about the collective organization of a team. Based on this idea, the aim of this chapter is to describe the network measurements that can provide a macro-perspective of the team's structure and dynamics.

Keywords Social network analysis · Graph theory · Digraphs · General measures · Macro-analysis

6.1 Total Links

Definition 6.1 (Wasserman and Faust 1994) Given one unweighted graph G with n vertices. The total links index, L, of G is calculated as

$$L = \sum_{i=1}^{n} \sum_{\substack{j=1 \\ j>i}}^{n} a_{ij},$$

(6.1)

where a_{ij} are elements of the adjacency matrix of a G.

Definition 6.2 (Rubinov and Sporns 2010) Given one weighted graph G with n vertices. The total links index, L^w, of G is obtained by

$$L^w = \sum_{i=1}^{n} \sum_{\substack{j=1 \\ j>i}}^{n} a_{ij},$$

(6.2)

where a_{ij} are elements of the weighted adjacency matrix of a G.

F.M. Clemente et al., *Social Network Analysis Applied to Team Sports Analysis*,
SpringerBriefs in Applied Sciences and Technology,
DOI 10.1007/978-3-319-25855-3_6

Definition 6.3 (Wasserman and Faust 1994) Given one unweighted digraph G with n vertices. The total links index, L_D, of G is determined by

$$L_D = \sum_{i=1}^{n} \sum_{\substack{j=1 \\ j \neq i}}^{n} a_{ij}, \tag{6.3}$$

where a_{ij} are elements of the adjacency matrix of a G.

Definition 6.4 (Rubinov and Sporns 2010) Given one weighted digraph G with n vertices. The total links index, L_D^w, of G is calculated as

$$L_D^w = \sum_{i=1}^{n} \sum_{\substack{j=1 \\ j \neq i}}^{n} a_{ij}, \tag{6.4}$$

where a_{ij} are elements of the weighted adjacency matrix of a G.

6.1.1   Team Sports Network Interpretation

The Total Links measure it is the absolute number of the total interactions conducted between teammates during the match. Thus, a higher than average Total Links index of a team is an indicator of strong cooperation between team players. This index may also be correlated with a high probability of the players to interact successfully with one another, which may result in long ball possession, good performance, and generally strong collective organization against the opponent team.

6.2   Network Density

Definition 6.5 (Pavlopoulos et al. 2011) Given one unweighted graph G with n vertices. The density index, Δ, of G is calculated as

$$\Delta = \frac{2 \times L}{n(n-1)}. \tag{6.5}$$

where L is the total links index of a G.

Definition 6.6 (Wasserman and Faust 1994) Given one weighted graph G with n vertices. The density index, Δ^w, of G is calculated as

$$\Delta^w = \frac{2 \times L^w}{n(n-1)}. \tag{6.6}$$

where L^W is the total links index of a G.

Definition 6.7 (Pavlopoulos et al. 2011) Given one unweighted digraph G with n vertices. The density index, Δ_D, of G is calculated as

$$\Delta_D = \frac{L_D}{n(n-1)}. \tag{6.7}$$

where L_D is the total links index of a G.

Definition 6.8 (Wasserman and Faust 1994) Given one weighted digraph G with n vertices. The density index, Δ_D^w, of G is calculated as

$$\Delta_D^w = \frac{L_D^w}{n(n-1)}. \tag{6.8}$$

where L_D^w is the total links index of a G.

6.2.1 Team Sports Network Interpretation

The density of team network is a relative index that also measures the overall affection between teammates. In graph theory, the density of a (directed) graph is the proportion of the maximum possible links present between nodes.

6.3 Distance

Definition 6.9 (Pavlopoulos et al. 2011) Given one unweighted graph G with n vertices. The average distance index, $\bar{d}$, of G is determined by

$$\bar{d} = \frac{2}{n(n-1)} \sum_{i=1}^{n} \sum_{\substack{j=1 \\ j>i}}^{n} d(n_i, n_j), \tag{6.9}$$

where $d(n_i, n_j)$ is a geodesic distance between n_i and n_j.

Remark 6.1 (Pavlopoulos et al. 2011) If $d(n_i, n_j) = \infty$, then we consider that $d(n_i, n_j) = \max(d(n_i, n_j) + 1)$.

Definition 6.10 (Wasserman and Faust 1994) Given one unweighted digraph G with n vertices. The average distance index, $\bar{d}$, of G is determined by

$$\bar{d} = \frac{2}{n(n-1)} \sum_{i=1}^{n} \sum_{\substack{j=1 \\ j \neq i}}^{n} d(n_i, n_j), \tag{6.10}$$

where $d(n_i, n_j)$ is a geodesic distance between n_i and n_j.

Definition 6.11 (Pavlopoulos et al. 2011) Given one weighted graph G with n vertices. The average distance index, $\overline{d^w}$, of G is determined by

$$\overline{d^w} = \frac{2}{n(n-1)} \sum_{i=1}^{n} \sum_{\substack{j=1 \\ j > i}}^{n} d^w(n_i, n_j), \tag{6.11}$$

where $d^w(n_i, n_j)$ is a geodesic distance between n_i and n_j.

Remark 6.2 If $d^w(n_i, n_j) = \infty$, then we consider that $d^w(n_i, n_j) = \max(d^w(n_i, n_j) + 1)$.

Definition 6.12 (Wasserman and Faust 1994) Given one weighted digraph G with n vertices. The average distance index, $\overline{d^w}$, of G is determined by

$$\overline{d^w} = \frac{2}{n(n-1)} \sum_{i=1}^{n} \sum_{\substack{j=1 \\ j \neq i}}^{n} d^w(n_i, n_j), \tag{6.12}$$

where $d^w(n_i, n_j)$ is a geodesic distance between n_i and n_j.

6.3.1   Team Sports Network Interpretation

The distances between players in a network may be an important macro-characteristic of the network as a whole (Hanneman and Riddle 2005). If distances are great, it may suggest that the ball take a long time to move for the teammates. It may also be that some players are out of playing of, and influenced by others. The variability across the players in the distances that they have from other players may be a basis for differentiation and even stratification (Hanneman and Riddle 2005). The players who are closer to more others may be able to exert more power than those who are more distant.

6.4 Network Diameter

Definition 6.13 (Wasserman and Faust 1994) Given one unweighted graph G with n vertices. The diameter index, D, of G is calculated as

$$D = \max_{i,j} d(n_i, n_j), \tag{6.13}$$

where $d(n_i, n_j)$ is a geodesic distance between n_i and n_j.

Remark 6.3 The diameter index of unweighted digraph is determined the similar form that unweighted graph.

Definition 6.14 (Wasserman and Faust 1994) Given one unweighted graph G with n vertices. The diameter index, D^w, of G is calculated as

$$D^w = \max_{i,j} d^w(n_i, n_j), \tag{6.14}$$

where $d^w(n_i, n_j)$ is a geodesic distance between n_i and n_j.

Remark 6.4 The diameter index of weighted digraph is determined the similar form that weighted graph.

6.4.1 Team Sports Network Interpretation

The diameter of a graph is related to the distance between players. In graph theory, two players are connected if a sequence of players exists and their connections are adjacent. The diameter of a graph is the maximum distance (the length of the largest geodesic) between any two connected players.

6.5 Clique

Definition 6.15 (Pavlopoulos et al. 2011) A clique in an unweighted graph is a subgraph G_S which is complete.

Definition 6.16 (Wasserman and Faust 1994) Let G be unweighted graph with n vertices. The p-clique is a subgraph with vertex set V_s ($V_s \subseteq V$), such that

$$d(n_i, n_j) \leq p \tag{6.15}$$

for all $n_i, n_j \in V_s$.

Remark 6.5 (Wasserman and Faust 1994) The size of p-clique is the number of vertices it contains.

Definition 6.17 (Wasserman and Faust 1994) Let G be unweighted digraph with n vertices.

(i)   A weakly connected p-clique is a subgraph in which all vertices are weakly n-connected, and there are no additional vertices that are also weakly n-connected to all vertices in the subgraph.

(ii) A unilaterally connected p-clique is a subgraph in which all vertices are unilaterally n-connected, and there are no additional vertices that are also unilaterally n-connected to all vertices in the subgraph.

(iii) A strongly connected p-clique is a subgraph in which all vertices are strongly n-connected, and there are no additional vertices that are also strongly n-connected to all vertices in the subgraph.

(iv) A recursively connected p-clique is a subgraph in which all vertices are recursively n-connected, and there are no additional vertices that are also recursively n-connected to all vertices in the subgraph.

Definition 6.18 (Wasserman and Faust 1994) Let G be weighted graph with n vertices. A subgraph G_S with vertex set V_s ($V_s \subseteq V$), is a clique at level c if for all $n_i, n_j \in V_s$ such that

$$w_{ij} \geq c \qquad (6.16)$$

And there is no vertex n_k, such that

$$w_{ki} \geq c \qquad (6.17.)$$

for all $n_i \in V_s$.

Definition 6.19 (Fagiolo 2007; Rubinov and Sporns 2010) Let G be unweighted graph with n vertices. The clustering coefficient index, $\overline{CL}$, of G is calculated by

$$\overline{CL} = \frac{1}{n} \sum_{i=1}^{n} CL(n_i). \qquad (6.18)$$

where $CL(n_i)$ is the clustering coefficient index of a vertex n_i.

Definition 6.20 (Fagiolo 2007; Rubinov and Sporns 2010) Let G be unweighted digraph with n vertices. The clustering coefficient index, $\overline{CL_D}$, of G is calculated by

$$\overline{CL_D} = \frac{1}{n} \sum_{i=1}^{n} CL_D(n_i). \qquad (6.19)$$

where $CL_D(n_i)$ is the clustering coefficient index of a vertex n_i.

Definition 6.21 (Fagiolo 2007; Rubinov and Sporns 2010) Let G be weighted graph with n vertices. The clustering coefficient index, $\overline{CL^w}$, of G is calculated by

$$\overline{CL^w} = \frac{1}{n} \sum_{i=1}^{n} CL^w(n_i). \tag{6.20}$$

where $CL^w(n_i)$ is the clustering coefficient index of a vertex n_i.

Definition 6.22 (Fagiolo 2007; Rubinov and Sporns 2010) Let G be weighted digraph with n vertices. The clustering coefficient index, $\overline{CL_D^w}$, of G is calculated by

$$\overline{CL_D^w} = \frac{1}{n} \sum_{i=1}^{n} CL_D^w(n_i). \tag{6.21}$$

where $CL_D^w(n_i)$ is the clustering coefficient index of a vertex n_i.

6.5.1 Team Sports Network Interpretation

The size of a clique comes from the number of vertices it contains (Pavlopoulos et al. 2011). It can used to identify the frequency of specific patterns of interactions between teammates. In the case, will be possible to identify the triangulations that emerge in match and the dyadic relationships that occurs and measure the regularity of these connections.

6.6 Network Heterogeneity

Definition 6.23 (Horvath 2011) Let G be unweighted graph with n vertices. The heterogeneity index, H, is calculated as

$$H = \sqrt{\frac{n \sum k_i^2 - \left(\sum k_i\right)^2}{\left(\sum k_i\right)^2}}, \tag{6.22}$$

with $i = 1, \ldots, n$ and k_i is degree centrality index of vertex n_i.

Definition 6.24 (Estrada and Hatano 2010) Let G be unweighted graph with n vertices. The variance of vertex degrees index, VAR, is calculated as

$$VAR = \frac{1}{n} \sum_{i=1}^{n} \left(k_i - \bar{k}\right)^2, \tag{6.23}$$

with k_i is degree centrality index of vertex n_i and $\bar{k}$ is the average of degrees.

Remark 6.6 In digraphs, weighted digraphs and weighted graphs, the variance of vertex degrees index is determined the similar form that unweighted graphs.

Definition 6.25 Let G be unweighted graph with n vertices. The variation coefficient of vertex degrees index, *VAR*, is calculated as

$$CVD = \frac{\sqrt{VAR}}{\bar{k}},\qquad(6.24)$$

with *VAR* is variance of vertex degrees index and $\bar{k}$ is the average of degrees.

Remark 6.7 In digraphs, weighted digraphs and weighted graphs, the variation coefficient of vertex degrees index is determined the similar form that unweighted graphs.

Definition 6.26 (Estrada and Hatano 2010) Let G be unweighted graph with n vertices. The normalized heterogeneity index, H_ρ, is calculated as

$$H_\rho = \frac{1}{n - 2\sqrt{n-1}} \sum_{i,j=1}^{n} \left\{ \frac{1}{k_i} + \frac{1}{k_j} - \frac{2}{\sqrt{k_i k_j}} \right\},\qquad(6.25)$$

with k_i is degree centrality index of vertex n_i and $(n_i, n_j) \in E$.

Remark 6.8 In weighted graphs, the normalized heterogeneity index is determined the similar form that unweighted graphs.

Definition 6.27 (Wilson et al. 2013) Let G be unweighted digraph with n vertices. The normalized heterogeneity index, H_{D_ρ}, is calculated as

$$H_{D_\rho} = \frac{1}{n - 2\sqrt{n-1}} \sum_{i,j=1}^{n} \left\{ \frac{1}{k_i^{out}} + \frac{1}{k_j^{in}} - \frac{2}{\sqrt{k_i^{out} k_j^{in}}} \right\},\qquad(6.26)$$

with k_i^{out} is outdegree index of vertex n_i, k_j^{in} is indegree index of vertex n_j and $(n_i, n_j) \in E$.

Remark 6.9 In weighted digraphs, the normalized heterogeneity index is determined the similar form that unweighted graphs.

6.6.1   Team Sports Network Interpretation

The heterogeneity can be used to measure the variation of connectivity across the players. Many complex networks have been found to exhibit an approximate scale-free topology, which implies that these networks are very heterogeneous. In the case of football analysis, greater values of heterogeneity reveal a non-cohesion interactional process between teammates.

6.7 Transitivity

Definition 6.28 (Rubinov and Sporns 2010) Let G be unweighted graph with n vertices. The transitivity index, T, is calculated as

$$T = \frac{\sum_i \left[\sum_{h \neq i,j} a_{ij} a_{ih} a_{jh} \right]}{\sum_i k_i (k_i - 1)} = \frac{\sum_{i=1}^{n} (A^3)_{ii}}{\sum_i k_i (k_i - 1)} \tag{6.27.}$$

with a_{ij} are elements of the adjacency matrix of a G and k_i is degree centrality index of vertex n_i.

Definition 6.29 (Rubinov and Sporns 2010) Let G be unweighted digraph with n vertices. The transitivity index, T_D, of is calculated as

$$T_D = \frac{\sum_i \left[\sum_{j,h} \left(a_{ij} + a_{ji} \right) \left(a_{ih} + a_{hi} \right) \left(a_{jh} + a_{hj} \right) \right]}{2 \sum_i \left[(k_i^{out} + k_i^{in})(k_i^{out} + k_i^{in} - 1) - 2 \sum_{j \neq i} a_{ij} a_{ji} \right]} \tag{6.28}$$

$$= \frac{(A + A^T)_{ij}^3}{2 \sum_i \left[(k_i^{out} + k_i^{in})(k_i^{out} + k_i^{in} - 1) - 2 \sum_{j \neq i} a_{ij} a_{ji} \right]}$$

with a_{ij} are elements of the adjacency matrix of a G and k_i^{out} is outdegree index of vertex n_i and k_i^{in} is indegree index of vertex n_i.

Definition 6.30 (Rubinov and Sporns 2010) Let G be weighted graph with n vertices. The transitivity index, T^w, is calculated as

$$T^w = \frac{\sum_i \left[\sum_{h \neq i,j} a_{ij}^{\frac{1}{3}} a_{ih}^{\frac{1}{3}} a_{jh}^{\frac{1}{3}} \right]}{\sum_i k_i (k_i - 1)} = \frac{\sum_{i=1}^{n} \left[\left(A^{[\frac{1}{3}]} \right)_{ii}^3 \right]}{\sum_{i=1}^{n} k_i (k_i - 1)} \tag{6.29}$$

with a_{ij} are elements of the weighted adjacency matrix of a G, k_i is degree centrality index of vertex n_i and $A^{[\frac{1}{3}]} = \left[a_{ij}^{\frac{1}{3}} \right]$, thus the matrix obtained from A by taking the 3rd root of each entry.

Definition 6.31 (Rubinov and Sporns 2010) Let G be weighted digraph with n vertices. The transitivity index, T_D^w, is calculated as

$$T_D^w = \frac{\sum_i \left[\sum_{j,h} \left(a_{ij}^{\frac{1}{3}} + a_{ji}^{\frac{1}{3}} \right) \left(a_{ih}^{\frac{1}{3}} + a_{hi}^{\frac{1}{3}} \right) \left(a_{jh}^{\frac{1}{3}} + a_{hj}^{\frac{1}{3}} \right) \right]}{2 \sum_i \left[(k_i^{out} + k_i^{in})(k_i^{out} + k_i^{in} - 1) - 2 \sum_{j \neq i} a_{ij} a_{ji} \right]} \tag{6.30}$$

$$= \frac{\sum_i \left[\left[A^{\left[\frac{1}{3}\right]} + (A^T)^{\left[\frac{1}{3}\right]} \right]_{ii} \right]^3}{2 \sum_i \left[(k_i^{out} + k_i^{in})(k_i^{out} + k_i^{in} - 1) - 2 \sum_{j \neq i} a_{ij} a_{ji} \right]}$$

with a_{ij} are elements of the weighted adjacency matrix of a G, k_i^{out} is outdegree index of vertex n_i, k_i^{in} is indegree index of vertex n_i and $A^{\left[\frac{1}{3}\right]} = \left[a_{ij}^{\frac{1}{3}} \right]$.

6.7.1 Team Sports Network Interpretation

This measure allows to identify balanced triads and to identify the "equilibrium" or natural state toward which triadic relationships tend. Transitivity allows identifying the capacity to the triad of players act in a balance way and not with tendencies such as pass for the same player.

6.8 Reciprocity

Definition 6.32 (Garlaschelli and Loffredo 2004) Let G be unweighted digraph with n vertices. The reciprocity index, R_D, of G is calculated as

$$R_D = \frac{L^{\leftrightarrow}}{L} \tag{6.31}$$

with $L^{\leftrightarrow}$ is the number of links pointing in both directions and L is the total links.

Definition 6.33 (Squartini et al. 2013) Let G be weighted digraph with n vertices. The weighted reciprocity index, R_D^w, of G is obtained by

$$R_D^w = \frac{\sum_i \sum_{j \neq i} w_{ij}^{\leftrightarrow}}{\sum_i \sum_{j \neq i} w_{ij}} \tag{6.32}$$

with $w_{ij}^{\leftrightarrow} \equiv min\left[w_{ij}, w_{ji}\right]$ and w_{ij} that is the weight of the $(n_i, n_j) \in E$.

6.8.1 Team Sports Network Interpretation

Reciprocity measures the tendency of players' pairs to form mutual connections between each other. In some cases it is visible an equilibrium tendency toward dyadic relationships to be either null or reciprocated, and that asymmetric ties may be unstable. A network that has a predominance of null or reciprocated ties over asymmetric connections may be a more "equal" or "stable" network than one with a predominance of asymmetric connections (which might be more of a hierarchy).

6.9 Global Centralization

Definition 6.34 (Pavlopoulos et al. 2011) Let G be unweighted graph with n vertices. The group degree centralization index, GC_D, of G is obtained by

$$GC_D = \frac{\sum_{i=1}^{n}[C_D(n^*) - C_D(n_i)]}{[(n-2)(n-1)]} \qquad (6.33)$$

with $C_D(n^*)$ is the largest degree of a vertex $n^* \in V$.

Remark 6.10 The group degree centralization index is determined the similar form in the cases unweighted digraphs, weighted graphs and weighted digraphs.

Definition 6.35 (Wasserman and Faust 1994) Let G be unweighted graph with n vertices. The coefficient of variation of group degree index, $CVGC_D$, of G is obtained by

$$CVGC_D = \frac{\sqrt{S_D^2}}{\overline{C_D}} = \frac{\sqrt{\left[\sum_{i=1}^{n}\left(C_D(n_i) - \overline{C_D}\right)^2\right]/n}}{\sum_{i=1}^{n} C_D(n_i)/n}. \qquad (6.34)$$

Remark 6.11 The coefficient of variation of group degree index is determined the similar form in the cases unweighted digraphs, weighted graphs and weighted digraphs.

Definition 6.36 (Wasserman and Faust 1994) Let G be unweighted graph with n vertices. The group closeness centralization index, GC_C, of G is obtained by

$$GC_C = \frac{\sum_{i=1}^{n}\left[C'_{(C)}(n^*) - C'_{(C)}(n_i)\right]}{[(n-2)(n-1)]/(2n-3)} \qquad (6.35)$$

with $C'_{(C)}(n^*)$ is the largest standardized closeness of a vertex $n^* \in V$.

Remark 6.12 The group closeness centralization index is determined the similar form in the cases unweighted digraphs, weighted graphs and weighted digraphs.

Definition 6.37 (Wasserman and Faust 1994) Let G be unweighted graph with n vertices. The coefficient of variation of group closeness index, $CVGC_c$, of G is obtained by

$$CVGC_c = \frac{\sqrt{S_C^2}}{C_{(C)}} = \frac{\sqrt{\left[\sum_{i=1}^{n}\left(C'_{(C)}(n_i) - C_{(C)}\right)^2\right]/n}}{\sum_{i=1}^{n} C'_{(C)}(n_i)/n}. \qquad (6.36)$$

Remark 6.13 The coefficient of variation of group closeness index is determined the similar form in the cases unweighted digraphs, weighted graphs and weighted digraphs.

Definition 6.38 (Wasserman and Faust 1994) Let G be unweighted graph with n vertices. The group betweenness centralization index, GC_b, of G is obtained by

$$GC_b = \frac{\sum_{i=1}^{n}\left[C_b'(n^*) - C_b'(n_i)\right]}{(n-1)} \tag{6.37}$$

with $C_b'(n^*)$ is the largest standardized betweenness of a vertex $n^* \in V$.

Remark 6.14 The group betweenness centralization index is determined the similar form in the cases unweighted digraphs, weighted graphs and weighted digraphs.

Definition 6.39 (Wasserman and Faust 1994) Let G be unweighted graph with n vertices. The coefficient of variation of group betweenness index, $CVGC_b$, of G is obtained by

$$CVGC_b = \frac{\sqrt{S_b^2}}{C_b} = \frac{\sqrt{\left[\sum_{i=1}^{n}\left(C_b'(n_i) - \overline{C_b}\right)^2\right]/n}}{\sum_{i=1}^{n}C_b'(n_i)/n}. \tag{6.38}$$

Remark 6.15 The coefficient of variation of group betweenness index is determined the similar form in the cases of unweighted digraphs, weighted graphs and weighted digraphs.

6.9.1 Team Sports Network Interpretation

Centralization is the measurement that shows whether a network has a star-like topology or whether the nodes of the network have on average the same connectivity. The closer the centralization is to 1, the more likely is the network to have a star-like topology, thus a tendency to play for the same player. The closer to 0, the more likely it is that the nodes of the network have on average the same connectivity, thus representing a more homogenous type of interaction.

6.10 Global Prestige

Definition 6.40 (Wasserman and Faust 1994) Let n_i be a vertex of unweighted digraph G with n vertices. The average proximity prestige index, P_P, variance proximity prestige index S_P^2 and variation coefficient proximity prestige index are determined by, respectively,

$$\overline{P_P} = \sum_{i=1}^{n} \frac{P_P(n_i)}{n},$$

(6.39)

$$S_P^2 = \sum_{i=1}^{n} \frac{\left(P_P(n_i) - \overline{P_P}\right)^2}{n},$$

(6.40)

$$VC_P = \frac{\sqrt{S_P^2}}{\overline{P_P}}$$

(6.41)

where $P_P(n_i)$ is the proximity prestige index of the vertex n_i.

Remark 6.16 In weighted digraph the average proximity prestige index, variance proximity prestige index and variation coefficient proximity prestige index are determined the similar form that unweighted digraphs.

6.10.1   Team Sports Network Interpretation

The global prestige allows identifying the specific value of participation of the team during the match. This value make possible to characterize the specific properties of the team into be prominent during the match.

References

Estrada, E., & Hatano, N. (2010). Resistance distance, information centrality, node vulnerability and vibrations in complex networks. *Network Science* (pp. 13–29). London: Springer.

Fagiolo, G. (2007). Clustering in complex directed networks. *Physical Review E, 76*(2), 026107.

Garlaschelli, D., & Loffredo, M. I. (2004). Patterns of link reciprocity in directed networks. *Physical Review Letters, 93*(26), 268701.

Hanneman, R. A., & Riddle, M. (2005). *Introduction to Social Network Methods.* Riverside, CA, USA: University of California, Riverside.

Horvath, S. (2011). *Weighted Network Analysis: Applications in Genomics and Systems Biology.* New York: Springer.

Pavlopoulos, G. A., et al. (2011). Using graph theory to analyze biological networks. *BioData Mining, 4*(1), 10.

Rubinov, M., & Sporns, O. (2010). Complex network measures of brain connectivity: Uses and interpretations. *NeuroImage, 52*(3), 1059–1069.

Squartini, T. et al. (2013). Reciprocity of weighted networks. *Scientific Reports, 3*(2729).

Wasserman, S., & Faust, K. (1994). *Social Network Analysis: Methods and Applications.* New York, USA: Cambridge University Press.

Wilson, R., et al. (2013). *Computer Analysis of Images and Patterns.* New York, USA: Springer.

Chapter 7
Argentina's Network Analysis in FIFA World Cup 2014: A Case Study

Abstract This chapter aims to identify how digraph analysis can be used in the scientific analysis of team sports. Therefore, a case study of Argentina's football team during FIFA World Cup 2014 was carried out. The one-way ANOVA tested the variance between tactical positions in the centrality measures of players. Another analysis of variance was made to test the general properties of graphs in different phases of tournament. Finally, the relationship between general measurements and team's performance was analysed. A total of 7 matches from the Argentina during FIFA World Cup 2014 tournament were analyzed and codified in this case study. The results found statistical differences between tactical positions in centrality measurements. No statistical differences were found between phases of tournament in general measurements. It were found significance correlations of shots with total links and density.

Keywords Social network analysis · Soccer · Match analysis · Case-study

7.1 Methods

7.1.1 Sample

A total of 7 matches from the Argentina during FIFA World Cup 2014 tournament were analyzed and codified in this case study. A total of 2520 passes between teammates were recorded and processed. A total of 14 weighted adjacency matrices (each per half) and corresponding digraphs were generated and used to compute the centrality metrics.

© The Author(s) 2016
F.M. Clemente et al., *Social Network Analysis Applied to Team Sports Analysis*,
SpringerBriefs in Applied Sciences and Technology,
DOI 10.1007/978-3-319-25855-3_7

7.1.2 Data Collecting and Processing

Argentina's matches in FIFA World Cup 2014 were examined. The players of national team were codified by their tactical position on the basis of the tactical lineup of Argentina's team. Three football coaches with more than five years of experience, to ensure the reliability of decision made, classified the tactical lineup of the team. During the matches, the tactical lineup of some teams changed, and in these situations, the national teams were classified on the basis of tactical lineup on which each team spent more time. The reliability of tactical line-up and tactical positions classifications were measured in a test-retest procedure (Robinson and O'Donoghue 2007). The Cohen's Kappa test adhering to a 30 day interval was used. A Kappa value of 0.92 was obtained after testing the full data (tactical lineup), thus ensuring a recommended margin for this type of procedures (Robinson and O'Donoghue 2007).

The tactical lineup was used in codifying the tactical position of each player. A techno-tactical assignment was adopted to positional roles (Di Salvo et al. 2007), and the tactical position of goalkeeper was added. The tactical assignment can be verified in Fig. 7.1.

The linkage indicator of passes made between teammates was defined for this study. Therefore, all attacking instants with more than one pass were observed. Each sequence of passes was classified as a unit of attack (Clemente et al. 2015a, b). An attacking unit started at the moment that a team player made a successful pass to a teammate and finished when the team lost possession of the ball (e.g., ball out of boundaries, ball out of shot, and unsuccessful pass to a teammate) (Clemente et al. 2015a, b). An adjacency matrix was generated per unit of attack. This matrix represents the connections between a node (player) and an adjacency node (teammate) (Passos et al. 2011). Each pass between nodes was codified as 1 (one),

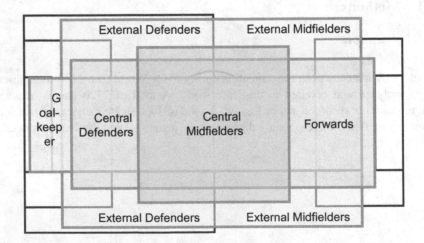

Fig. 7.1 Tactical positions codified based on match analysis

and no passes between teammates were codified as 0 (zero). More than one pass between the same nodes were codified with the number of passes (Clemente et al. 2015a, b). Each player was classified with a number between P1 and P11 for easy codification.

The final weighted adjacency matrix that comprised the sum of both weighted matrices (1st half and 2nd half) from all units of attack was computed at the end of each match. The metrics were computed on the basis of the final weighted adjacency matrix. The observation and codification of sequences of passes were performed by the same researcher with more than five years of experience in match analysis to minimize inter-reliability error. The observer was previously trained and was tested in a test-retest procedure to ensure the reliability of data (Clemente et al. 2015a, b). Cohen's Kappa test adhered to a 30-day interval for re-analysis to avoid task familiarity issues (Robinson and O'Donoghue 2007). A Kappa value of 0.76 was obtained after testing 15 % of the full data. The Kappa value ensured a recommended margin for this type of procedures (Robinson and O'Donoghue 2007).

7.1.3 Variables of the Study

This case study aimed to analyze the variance of centrality measures between tactical positions and attempted to identify the most prominent tactical positions of Argentina's team that contribute to building the attack. Moreover, the general properties of Argentina's team were also analyzed between different moments during competition. In the case of centrality measures, the tactical position was defined as independent variable. In the case of general properties of digraphs, the phases of tournament was defined as independent variable. To identify the most prominent tactical positions in all national teams, the social network analysis approach, particularly the network metrics of centrality, was used with the following considerations: (i) $P'^w_{(D-in)}(n_i)$; (ii) $C'^w_{(D-out)}(n_i)$; (iii) $C'^w_{(C)}(n_i)$; and (iv) $C'_b(n_k)$. Those values were used as dependent variables in this study. In the case of general properties, the following dependent variables were used: (i) L^w_D; (ii) Δ^w_D; (iii) $CL^w_D(n_i)$; and (iv) D^w.

7.1.4 Statistical Procedures

The influences of tactical position on the micro measures and the phases of competition in macro measures were analyzed using one-way ANOVA after validating the normality and homogeneity assumptions. The assumption of normality for each univariate-dependent variable was examined using Kolmogorov–Smirnov tests (p-value > 0.05). The assumption of the homogeneity of the variance/covariance matrix of each group was examined using the Box's M Test (Pallant 2011). When the ANOVA detected significant statistical differences between factors it was tested the post hoc by using Tukey's HSD (O'Donoghue 2012). The following scale was used to classify the effect size (partial eta square) of the test (Pierce et al. 2004): small, 0.14–0.36; moderate, 0.37–0.50; large, 0.51–1. All statistical analyses were performed using IBM SPSS Statistics (version 21) at a significance level of $p < 0.05$.

The relationship between general measurements and team performance variables (goals scored, shots, and shots on goal) was investigated using Pearson product moment correlation coefficient. Preliminary analysis was performed to ensure no violation of the assumptions of normality, linearity, and homoscedasticity (Pallant 2011). The following scales were used to classify the correlation strength (Hopkins et al. 1996): very small, 0–0.1; small, 0.1–0.3; moderate, 0.3–0.5; large, 0.5–0.7; very large, 0.7–0.9; 0.9–1, nearly perfect; 1, perfect.

7.2   Results

The analysis of variance to centrality measurements between tactical positions were analyzed in this case study. The descriptive statistics can be found in the following Table 7.1.

One-way ANOVA revealed statistical differences between tactical positions in Degree Prestige ($F = 19.899$; $p = 0.001$; *Effect Size* = 0.584; *Large Effect Size*), Degree Centrality ($F = 12.155$; $p = 0.001$; *Effect Size* = 0.461; *Moderate Effect Size*), Closeness Centrality ($F = 10.211$; $p = 0.001$; *Effect Size* = 0.418; *Moderate Effect Size*), and Betweenness Centrality ($F = 6.286$; $p = 0.001$; *Effect Size* = 0.307; *Small Effect Size*).

The analysis of variance to macro measurements between phases of tournament were also analyzed in this case study. The descriptive statistics can be found in the following Table 7.2.

One-way ANOVA revealed statistical differences between phases of tournament in Total Links ($F = 2.783$; $p = 0.281$; *Effect Size* = 0.848; *Large Effect Size*), Density ($F = 2.778$; $p = 0.282$; *Effect Size* = 0.847; *Large Effect Size*), clustering coefficient ($F = 0.716$; $p = 0.653$; *Effect Size* = 0.589; *Large Effect Size*), and diameter ($F = 0.571$; $p = 0.716$; *Effect Size* = 0.533; *Large Effect Size*).

The relationship between team performance variables (goals scored, overall shots, and shots on goal) and the characteristics of the network graphs (total arcs, network density, clustering coefficient, and diameter) was investigated using Pearson product-moment correlation coefficient. The values of the coefficients are shown in Table 7.3.

The shots showed a very large negative correlation with total links (r = −0.756; p = 0.049) and density (r = −0.756; p = 0.049). The remaining performance variables did not have significance correlations with general measurements.

Table 7.1 Descriptive statistics (mean and standard deviation) and post hoc values for dependent variables between tactical positions

	GK	ED	CD	CM	EM	FW
$P^w_{(D-in)}(n_i)$	1.97 (1.39)[b,c,d,e]	8.51 (1.92)[a,d]	8.92 (1.54)[a,d]	11.62 (3.31)[a,b,c,f]	11.30 (3.02)[d,e]	5.70 (2.28)[d,e]
$C^w_{(D-out)}(n_i)$	3.49 (1.82)[b,c,d,e]	9.93 (1.92)[a,f]	10.10 (1.64)[a,f]	11.60 (4.55)[a,f]	9.00 (3.40)[a,f]	3.67 (1.51)[b,c,d,e]
$C^w_{(D-out)}(n_i)$	7.74 (1.92)[b,c,d,e]	9.22 (0.66)[a,f]	9.29 (1.11)[a,f]	9.81 (0.65)[a,f]	9.06 (0.91)[a,f]	7.68 (0.78)[b,c,d,e]
$C_b(n_k)$	1.74 (1.22)[b,c,d]	10.16 (4.63)[a]	11.32 (6.45)[a]	12.20 (6.49)[a,e]	6.68 (3.22)[d]	5.34 (4.17)[d]

[a] Significantly different compared with goalkeeper (GK)
[b] External defenders (ED)
[c] Central defenders (CD)
[d] Central midfielders (CM)
[e] External midfielders (EM)
[f] Forwards (FW) at $p < 0.05$

Table 7.2 Descriptive statistics (mean and standard deviation) and post hoc values for dependent variables between phases of tournament

	Stage group	Round of 16	Quarter finals	Semi finals	Final
L_D^w	84.67 (2.89)	75.00 (0.00)	81.00 (0.00)	87.00 (0.00)	82.00 (0.00)
Δ_D^w	0.77 (0.03)	0.68 (0.00)	0.74 (0.00)	0.79 (0.00)	0.75 (0.00)
$CL_D^w(n_i)$	0.80 (0.04)	0.75 (0.00)	0.77 (0.00)	0.79 (0.00)	0.75 (0.00)
D^w	2.33 (0.58)	3.00 (0.00)	2.00 (0.00)	2.00 (0.00)	2.00 (0.00)

[a]Significantly different compared with stage group
[b]Round of 16
[c]Quarter finals
[d]Semi finals
[e]Final at $p < 0.05$

Table 7.3 Correlation values between the team performance variables and the network values provided by the metrics

	GS	GC	S	TA	ND	CC	D
Team attacking performance							
(1) GS: Goals scored	1	0.679	0.349	0.083	0.084	0.421	0.228
(2) GC: Goals ceded		1	−0.095	0.255	0.256	0.202	−0.062
(3) S: Shots			1	−0.756*	−0.756*	−0.336	0.456
Network performance							
(4) TL: L_D^w				1	1.000**	0.786*	−0.194
(5) ND: Δ_D^w					1	0.786*	−0.194
(6) CC: $CL_D^w(n_i)$						1	−0.362
(7) D: D^w							1

*Correlation is significant at p ≤ 0.050
**Correlation is significant at p = 0.001

7.3 Discussion

In this case study it was analyzed the Argentina's network during FIFA World Cup 2014. Centrality measurements and general properties were compared between tactical positions and phases of competition, respectively. The main results found statistical differences in centrality measurements between tactical positions. No statistical differences in general properties were found between phases of competition. Only performance variable of shots revealed significant correlations with total arcs and network density.

In the specific analysis of variance between tactical positions it was found that central midfielders and external midfielders had the greatest values in degree prestige. Previous studies revealed that midfielders are in general the key-elements of the team during attacking connections (Clemente et al. 2014, 2015a, b; Peña and

Touchette 2012; Duch et al. 2010; Malta and Travassos 2014). Nevertheless, previous studies that analyzed full sequences of play revealed that external defenders are the second most central players into receive the ball (Clemente et al. 2014, 2015a, b). In this specific case study it was found that external midfielder was the second central player into receive the ball. This finding may suggest that the team reveals a tendency to exploit the wings and to play using the width of the field ('large field') during attacking building (Costa et al. 2010). Another interesting finding was the small value of degree prestige associated with forward. This may reveal that during passing sequences the forward player is not a key element to receive the ball and the team opts by displace the ball from the center.

The analysis to degree centrality revealed that central midfielder and central defender had the greatest values of passes made for teammates. Previous studies in network analysis showed that external defenders, central defenders and central midfielders are the prominent players into pass the ball (Peña and Touchette 2012; Clemente et al. 2014, 2015a, b; Malta and Travassos 2014). Such results may suggest that ball circulation depends from the backward players to build the attack. In the specific case of Argentina, central defenders had greatest values of outdegree than external defenders, thus suggesting that the team opts to pass by the middle in the initial phase of attacking building and then exploit the wings of the field with passes to the external midfielders.

The analyses to the closeness centrality revealed that goalkeeper and forward were the farthest players in the team. This result suggests that the core of attacking play occurs in the middle of the field (length) and for that reason central defenders; external defenders and central midfielders had similar results between them. Similar results were found in previous studies (Clemente et al. 2015a, b).

Finishing the analysis to the centrality measurements it was found that central midfielder was the tactical position with greatest values in betweenness centrality. Goalkeeper, external midfielder and forward players had the smallest values. The midfielder acts as a link in modern football (Clemente et al. 2013). Moreover, in teams that opt to ball circulation, the prominence of midfielder is greater. For that reason, it is normal that greatest values of betweenness are associated with the specific position of midfielders.

The macro analysis to the digraph properties revealed no statistical differences between phases of tournament in Argentina's team. The results obtained may suggest that the pattern of play di not change from match to match. In previous studies it was found that higher levels of network density lead to better performance in football (Grund 2012). The same evidence it was found in the overall analysis performed in FIFA World Cup 2014 that analyzed the general properties between the teams that had the best performances and lowest performances (Clemente et al. 2015a, b). In the case of Argentina's team (2nd classified in the tournament) the values of total links ranched between 0.75 and 0.87. These values are in line with the best performances achieved during the tournament (Clemente et al. 2015a, b).

Greater values of total links may reveal a team that opts to play in ball circulation and attacking building. For that reason, teams such as Germany in FIFA World Cup 2014 and Spain in FIFA World Cup 2010 had greatest values in total arcs. The style of play influences this property. In teams that opt to play in counter attack or in attacking transition there are small values of total arcs because the players are not so connected. The total links also had very large negative correlations with shots. This suggests that greater values of total arcs lead to smaller values of shots. In fact, the teams that opt to ball circulation try to identify the best moment to attack and for that reason are more accurate and do not perform much shots. By other hand, teams that opt to play in direct attack try to score as faster as possible. Nevertheless, this is only a hypothesis that must be confirmed in further studies.

In other analysis, the density measure revealed a very large positive correlation with shots. This result may not confirm the possibility that teams that opt to play in ball circulation had small volume of shots. The density allows identifying the capacity of play evolving the majority of players. A team such as Barcelona (in tiki-taka's era) represents the exponent of network density (Peña and Touchette 2012). In the present study, only Argentina's team it was analyzed and for that reason it is not possible to confirm the tendencies revealed by statistical analysis.

Finally, no statistical correlations were found between clustering coefficient and diameter with the performance variables. Previous studies also revealed no statistical differences between these parameters between successful and unsuccessful teams during FIFA World Cup 2014 (Clemente et al. 2015a, b). For that reason, the two indicators may not be the better variables to discriminate the performances in football matches.

7.4 Practical Applications

In this study it was possible to identify the using centrality metrics it is possible to discriminate the prominent players in the team. Moreover, by crossing the coaches' analysis with results it is possible to analyze the style of play and the strategic options made by the team during matches. In other hand, the general properties of digraph may be better analyzed if crossed with meso analysis. For the case of practical applications it is possible to suggest that centrality metrics can reveal more important information to be used by coaches to optimize the training sessions and to make decisions during matches.

References

Clemente, F.M., et al. (2013, September). Activity profiles of soccer players during the 2010 World Cup. *Journal of Human Kinetics, 38*, 201–211.
Clemente, F. M., et al. (2014). Using network metrics to investigate football team players' connections: A pilot study. *Motriz, 20*(3), 262–271.
Clemente, F. M., et al. (2015a). General network analysis of national soccer teams in FIFA World Cup 2014. *International Journal of Performance Analysis in Sport, 15*(1), 80–96.

Clemente, F. M., et al. (2015b). Midfielder as the prominent participant in the building attack: A network analysis of national teams in FIFA World Cup 2014. *International Journal of Performance Analysis in Sport, 15*(2), 704–722.

Costa, I. T., et al. (2010). Influence of relative age effects and quality of tactical behaviour in the performance of youth football players. *International Journal of Performance Analysis in Sport, 10*(2), 82–97.

Di Salvo, V., et al. (2007). Performance characteristics according to playing position in elite soccer. *International Journal of Sports Medicine, 28*, 222–227.

Duch, J., Waitzman, J. S., & Amaral, L. A. (2010). Quantifying the performance of individual players in a team activity. *PLoS ONE, 5*(6), e10937.

Grund, T. U. (2012). Network structure and team performance: The case of English Premier League soccer teams. *Social Networks, 34*(4), 682–690.

Hopkins, K. D., Hopkins, B. R., & Glass, G. V. (1996). *Basic Statistics for the Behavioral Sciences*. Boston: Allyn and Bacon.

Malta, P., & Travassos, B. (2014). Characterization of the defense-attack transition of a soccer team. *Motricidade, 10*(1), 27–37.

O'Donoghue, P. (2012). *Statistics for Sport and Exercise Studies: An Introduction*. London and New York, UK and USA: Routledge, Taylor & Francis Group.

Pallant, J. (2011). *SPSS Survival Manual: A Step by Step Guide to Data Analysis Using the SPSS Program*. Australia: Allen & Unwin.

Passos, P., et al. (2011). Networks as a novel tool for studying team ball sports as complex social systems. *Journal of Science and Medicine in Sport, 14*(2), 170–176.

Peña, J.L., & Touchette, H. (2012). A network theory analysis of football strategies. In *arXiv Preprint arXiv* (p. 1206.6904).

Pierce, C. A., Block, R. A., & Aguinis, H. (2004). Cautionary note on reporting eta-squared values from multifactor ANOVA designs. *Educational and Psychological Measurement, 64*(6), 916–924.

Robinson, G., & O'Donoghue, P. (2007). A weighted kappa statistic for reliability testing in performance analysis of sport. *International Journal of Performance Analysis in Sport, 7*(1), 12–19.

Printed in the United States
By Bookmasters